Titles in This Series

765 **Robert Bieri and Ross Geoghegan,** Connectivity properties of group actions on non-positively curved spaces, 2003

764 **J. Spandaw,** Noether Lefschetz problems for degeneracy loci, 2003

763 **Yasuyuki Kachi and Eiichi Sato,** Segre's reflexivity and an inductive characterization os hyperquadrics, 2002

762 **Leiba Rodman, Ilya M. Spitkovsky, and Hugo Woerdeman,** Abstract band method via factorization, positive and band extensions of multivariable almost periodic matrix functions, and spectral estimation, 2002

761 **Oliver Druet and Emmanuel Hebey,** The AB program in geometric analysis : Sharp Sobolev inequalities and related problems, 2002

760 **Markus Banagl,** Extending intersection homology type invarients to non-Witt spaces, 2002

759 **Donald M. Davis,** From representation theory to homotopy groups, 2002

758 **Alan Forrest, John Hunton, and Johannes Kellendonk,** Topological invariants for projection method patterns, 2002

757 **Douglas Bowman,** q-difference operators, orthogonal polynomials, and symmetric expansions, 2002

756 **José Ignacio Cogolludo-Agustín,** Topological invariants of the complement to arrangements of rational plane curves, 2002

755 **M. A. Mandell and J. P. May,** Equivariant orthogonal spectra and S-modules, 2002

754 **Edward L. Green, Idun Reiten, and Øyvind Solberg,** Dualities on generalized Koszul algebras, 2002

753 **Daniel Panazzolo,** Desingularization of nilpotent singularities in families of planar vector fields, 2002

752 **Linus Kramer,** Homogeneous spaces, Tits buildings, and isoparametric hypersurfaces, 2002

751 **Bruce Allison, Georgia Benkart, and Yun Gao,** Lie algebras graded by the root systems BC_r, $r \geq 2$, 2002

750 **Masaki Izumi and Hideki Kosaki,** Kac algebras arising from composition of subfactors: General theory and classification, 2002

749 **Nanhua Xi,** The based ring of two-sided cells of affine Weyl groups of type $\widetilde{A}_{n-1}$, 2002

748 **Jürgen Ritter and Alfred Weiss,** The lifted root number conjecture and Iwasawa theory, 2002

747 **Armand Borel, Robert Friedman, and John W. Morgan,** Almost commuting elements in compact Lie groups, 2002

746 **Peter Niemann,** Some generalized Kac-Moody algebras with known root multiplicities, 2002

745 **Mikhail A. Lifshits and Werner Linde,** Approximation and entropy numbers of Volterra operators with application to Brownian motion, 2002

744 **Roger Chalkley,** Basic global relative invariants for homogeneous linear differential equations, 2002

743 **Heng Sun,** Spectral decomposition of a covering of $GL(r)$: the Borel case, 2002

742 **J. E. Gilbert, Y. S. Han, J. A. Hogan, J. D. Lakey, D. Weiland, and G. Weiss,** Smooth molecular functions and singular integral operators, 2002

741 **Francisco Santos,** Triangulations of oriented matroids, 2002

740 **Rick Durrett,** Mutual invadability implies coexistence in spatial models, 2002

739 **Georgios K. Alexopoulos,** Sub-Laplacians with drift on Lie groups of polynomial volume growth, 2002

Connectivity Properties of Group Actions on Non-Positively Curved Spaces

MEMOIRS
of the
American Mathematical Society

Number 765

Connectivity Properties of Group Actions on Non-Positively Curved Spaces

Robert Bieri
Ross Geoghegan

January 2003 • Volume 161 • Number 765 (second of 5 numbers) • ISSN 0065-9266

American Mathematical Society
Providence, Rhode Island

2000 *Mathematics Subject Classification.*
Primary 20F65, 53C23, 20C99, 20F69, 57N16; Secondary 20E08, 20J05, 20G25.

Library of Congress Cataloging-in-Publication Data

Bieri, Robert.

Connectivity properties of group actions on non-positively curved spaces / Robert Bieri, Ross Geoghegan.

p. cm. — (Memoirs of the American Mathematical Society, ISSN 0065-9266 ; no. 765)

"Volume 161, number 765 (second of 5 numbers)."

Includes bibliographical references.

ISBN 0-8218-3184-4 (alk. paper)

1. Geometric group theory. 2. Connections (Mathematics) 3. Global differential geometry. I. Geoghegan, Ross, 1963– II. Title. III. Series.

QA3 .A57 no. 765
[QA183]
510 s—dc21
[512′.2] 2002033242

Memoirs of the American Mathematical Society

This journal is devoted entirely to research in pure and applied mathematics.

Subscription information. The 2003 subscription begins with volume 161 and consists of six mailings, each containing one or more numbers. Subscription prices for 2003 are $555 list, $444 institutional member. A late charge of 10% of the subscription price will be imposed on orders received from nonmembers after January 1 of the subscription year. Subscribers outside the United States and India must pay a postage surcharge of $31; subscribers in India must pay a postage surcharge of $43. Expedited delivery to destinations in North America $35; elsewhere $130. Each number may be ordered separately; *please specify number* when ordering an individual number. For prices and titles of recently released numbers, see the New Publications sections of the *Notices of the American Mathematical Society.*

Back number information. For back issues see the *AMS Catalog of Publications.*

Subscriptions and orders should be addressed to the American Mathematical Society, P. O. Box 845904, Boston, MA 02284-5904, USA. *All orders must be accompanied by payment.* Other correspondence should be addressed to 201 Charles Street, Providence, RI 02904-2294, USA.

Memoirs of the American Mathematical Society is published bimonthly (each volume consisting usually of more than one number) by the American Mathematical Society at 201 Charles Street, Providence, RI 02904-2294, USA. Periodicals postage paid at Providence, RI. Postmaster: Send address changes to Memoirs, American Mathematical Society, 201 Charles Street, Providence, RI 02904-2294, USA.

This publication is indexed in *Science Citation Index*®, *SciSearch*®, *Research Alert*®, *CompuMath Citation Index*®, *Current Contents*®*/Physical, Chemical & Earth Sciences.*
Printed in the United States of America.

∞ The paper used in this book is acid-free and falls within the guidelines established to ensure permanence and durability.
Visit the AMS home page at http://www.ams.org/

10 9 8 7 6 5 4 3 2 1 08 07 06 05 04 03

Contents

Abstract. Generalizing the Bieri-Neumann-Strebel-Renz Invariants, this Memoir presents the foundations of a theory of (not necessarily discrete) actions ρ of a (suitable) group G by isometries on a proper CAT(0) space M. The passage from groups G to group actions ρ implies the introduction of "Sigma invariants" $\Sigma^k(\rho)$ to replace the previous $\Sigma^k(G)$ introduced by those authors. Their theory is now seen as a special case of what is studied here so that readers seeking a detailed treatment of their theory will find it included here as a special case.

We define and study "controlled k-connectedness (CC^k)" of ρ, both over M and over end points e in the "boundary at infinity" ∂M; $\Sigma^k(\rho)$ is by definition the set of all e over which the action is $(k-1)$-connected. A central theorem, the Boundary Criterion, says that $\Sigma^k(\rho) = \partial M$ if and only if ρ is CC^{k-1} over M. An Openness Theorem says that CC^k over M is an open condition on the space of isometric actions ρ of G on M. Another Openness Theorem says that $\Sigma^k(\rho)$ is an open subset of ∂M with respect to the Tits metric topology. When $\rho(G)$ is a discrete group of isometries the property CC^{k-1} is equivalent to $\ker(\rho)$ having the topological finiteness property "type F_k". More generally, if the orbits of the action are discrete, CC^{k-1} is equivalent to the point-stabilizers having type F_k. In particular, for $k = 2$ we are characterizing finite presentability of kernels and stabilizers.

Examples discussed include: locally rigid actions, translation actions on vector spaces (especially those by metabelian groups), actions on trees (including those of S-arithmetic groups on Bruhat-Tits trees), and SL_2 actions on the hyperbolic plane.

The first-named author was supported in part by a grant from the Deutsche Forschungsgemeinschaft.

The second-named author was supported in part by grants from the National Science Foundation.

1991 *Mathematics Subject Classification.* Primary: 20F65, 53C23, 20C99, 20F69, 57N16

Secondary: 20E08, 20J05, 20G25

Key words and phrases. CAT(0) space, Bieri-Neumann-Strebel invariant, controlled topology, finiteness properties, non-positively curved, isometric action

Received by the editor May 16, 2001.

Preface

This memoir presents a topological theory of (not necessarily discrete) actions of a group G by isometries on a proper CAT(0) metric space M. Examples of such spaces are numerous and well-known, perhaps the best known being simply connected manifolds of non-positive sectional curvature. Other examples include finite products of locally finite trees and, more generally, locally finite affine buildings.

For each such isometric action ρ of G on M and each $n \geq 0$ we define a subset $\Sigma^n(\rho)$ generalizing the invariants $\Sigma^n(G)$ of Bieri-Neumann-Strebel-Renz (abbreviated here to BNSR) [**BS 80**], [**BS 81**], [**BNS 87**], [**BRe 88**], [**Re 88**], [**Re 89**]. The case $n = 0$ is only relevant for non-cocompact actions and can be ignored in this preface. The case $n = 1$ can be sketched quickly and so we will concentrate on it first, noting that everything in the memoir is done for general n.

Here is a description of the "classical" $\Sigma^1(G)$, often called the Bieri-Neumann-Strebel Invariant of G. Assuming G finitely generated, let $h : \Gamma \to V$ be the canonical G-map from the Cayley graph (with respect to a chosen finite set of generators) to the G-vector space $V := G/G' \otimes_{\mathbb{Z}} \mathbb{R}$. Choose an inner product for V to make it a Euclidean space. Then each $e \in S(V)$, the sphere at infinity of V, defines half spaces in V: we say that $e \in \Sigma^1(G)$ if and only if the h-preimages of all these half spaces are connected subsets of Γ. Thus $\Sigma^1(G)$ is a subset of $S(V)$.

The Invariant $\Sigma^1(G)$ contains the information needed to decide whether a given subgroup S of G containing the commutator subgroup G' is finitely generated. If G is metabelian $\Sigma^1(G)$ is polyhedral and from knowledge of $\Sigma^1(G)$ one can read off whether or not G admits a finite presentation; indeed, conjecturally whether or not G is of type FP_m. The set $\Sigma^1(G)$ is also polyhedral when G is non-abelian and is the fundamental group of a compact 3-manifold M^3. In fact, in that case it has a description in terms of the unit sphere of the Thurston norm on $H_1(M^3)$.

In our generalization (sticking for now to $n = 1$) $\Sigma^1(\rho)$ is described similarly: one has a given action $\rho : G \to \mathrm{Isom}(M)$ of G on M and one chooses a G-map $h : \Gamma \to M$. The proper CAT(0) space M has a compactifying boundary ∂M and $\Sigma^1(\rho)$ is defined to be a subset of ∂M as follows: $e \in \Sigma^1(\rho)$ if and only if the h-preimages of horoballs in M centered at e are connected. Note that half-spaces in V, above, are horoballs, so this is a generalization. When the subgroup $\rho(G)$ is discrete (as is the case in the previous action on V) $\Sigma^1(\rho)$ is all of ∂M if and only if $\ker(\rho)$ is finitely generated.

We say that ρ is "controlled 0-connected" over e if and only if $e \in \Sigma^1(\rho)$. When n replaces 1, the relevant notion is called "controlled $(n-1)$-connected", i.e. preimages of horoballs at e are $(n-1)$-connected in an approximate sense, and we define $\Sigma^n(\rho)$ to be the set of those $e \in \partial M$ over which the action is controlled $(n-1)$-connected. If $\rho(G)$ is discrete in $\mathrm{Isom}(M)$ our Theorem A$'$ says that $\Sigma^n(\rho)$ is all of ∂M if and only if the kernel of the action has type F_n.

But one should not confine attention to actions where $\rho(G)$ is discrete. Rather, our viewpoint is that controlled $(n-1)$-connectedness over all of ∂M is an important invariant of arbitrary actions – one which happens to coincide with a finiteness property of kernels in the discrete case. Indeed two situations are interesting and are both discussed at length in the memoir: (1) the case where $\Sigma^n(\rho)$ is all of ∂M - what are its consequences? and (2) the case where $\Sigma^n(\rho)$ is a proper subset of ∂M - what kind of subset?

There is a related notion called "controlled $(n-1)$-connected over M": roughly, this means that the pre-images of balls in M are $(n-1)$-connected in an approximate but uniform sense. One of our theorems is the Boundary Criterion (see Chapter 10) that ρ is controlled $(n-1)$-connected over M if and only if ρ is controlled $(n-1)$-connected over every point of ∂M. This seems to be an important feature of the theory: an example where it is used in a crucial way is described in §10.7(B) where we study the natural action of $SL_2(\mathbb{Z}[1/m])$ on the hyperbolic plane.

Many of the ideas used in this memoir are present in some germinal form, explicit or implicit, in the earlier literature - that is, the literature on $\Sigma^n(G)$. Here are two features to note about the methods used here which might be considered novel:

1. Our point of view is that we are studying actions. Indeed, the earlier literature can be interpreted as studying translation actions of G on Euclidean spaces, but it is not usually presented that way. In that setting there is a "universal" translation action, namely the natural action of G by translations on the Euclidean space V defined above. In a sense, all such actions lie "under" this action. But there is no analogue of this universal action if we are discussing isometric actions of G on CAT(0) spaces.

2. In the translation case end points are always fixed. That is to say, if G acts by translations on a finite-dimensional Euclidean space, G acts trivially on the sphere at infinity. This need not be so for actions on CAT(0) spaces; for example, the action of $SL_2(\mathbb{Z})$ on the hyperbolic plane by Möbius transformations moves boundary points. The consequence for us is that there are not enough equivariant maps on the Cayley graph (and on its higher-dimensional analogues) to do what we want to do. Therefore we have to introduce a larger class of allowable moves called "finitary maps". These are new. We introduce "finitary maps" and the related notion of "sheaves of maps" in a self-contained essay in Chapter 4. These only involve elementary homotopy theoretical ideas, but there is some foundational material to be explained. The statements of most of our theorems can be understood without this material, but not the proofs (except for Theorem A - see below).

The logical arrangement of the memoir is that controlled connectivity over M is treated first. This is Part I. The boundary ∂M only appears in Part II where controlled connectivity over end points and the invariants $\Sigma^n(\rho)$ are discussed.

Readers knowledgeable about the BNSR-invariants $\Sigma^n(G)$ may want to go straight to Part II to see the generalizations $\Sigma^n(\rho)$. We have tried to make this possible: first read Chapter 4 on finitary maps and sheaves of maps (noting the introductory motivating paragraph §2.6) and then go to Part II.

To see precise statements of all our results go to Chapter 2 (for Part I) and Chapter 10 (for Part II). In fact we would urge any reader to begin by reading both of those sections.

Acknowledgments: Gaël Meigniez has independently obtained some of our main results in the special case $n = 1$ and also has some insight for $n > 1$. We have profited from discussions with him; in particular he pointed out that our proof of Theorem A$'$ actually proves the stronger Theorem A. Tom Farrell, Erik Pedersen and Kai-Uwe Bux also gave us useful insights.

CHAPTER 1

Introduction

A thorough outline of this memoir is given in Chapter 2 for Part I and in Chapter 10 for Part II. In this introduction we give a quick indication of what it is about.

1.1. Cocompact is an open condition

Let (M, d) be a simply connected proper metric space which is "non-positively curved" (i.e., CAT(0)). For example, M might be a complete simply connected Riemannian manifold of non-positive sectional curvature, or M might be a locally finite affine building. Let G be a group. We study actions of G on M by isometries. The space of all such actions, $\mathrm{Hom}(G, \mathrm{Isom}(M))$, carries the compact-open topology (with respect to the discrete topology on G and the compact-open topology on $\mathrm{Isom}(M)$). We emphasize that while G is discrete we are considering all actions by isometries, not just discrete actions.

An action $\rho : G \to \mathrm{Isom}(M)$ is *cocompact* if there is a compact subset K of M such that the G-translates of K cover M (i.e., $GK = M$). One of our results is

- Cocompactness is an open condition on ρ. In other words, the set of all cocompact actions ρ is open in $\mathrm{Hom}(G, \mathrm{Isom}(M))$.

1.2. Controlled connectivity

We do not prove this openness theorem directly. Rather it falls out naturally as a special case of more general considerations: it turns out that cocompactness can be viewed as the (-1)-dimensional case of an n-dimensional connectivity property of ρ which we call "controlled n-connected", abbreviated to "CC^n". Very roughly, the flavor of this is the following: Consider a free contractible G-CW-complex X and a G-map $h : X \to M$; then ρ is CC^n if the h-preimages of balls in M are, in some sense, coarsely n-connected subsets of X. This property is extracted and generalized from work on the "geometric invariants" $\Sigma^n(G)$ of the group G introduced in [**BS 80**], [**BNS 87**] [**BRe 88**] [**Re 88**], [**Re 89**]: these apply in the special case where $M = \mathbb{E}^m$ and ρ is an action by translations. The fact that cocompactness is an open condition is now the special case $n = 0$ of the following result:

- If G has type[1] F_n, CC^{n-1} is an open condition. See Theorem B.

There is a second special case where CC^{n-1} has a familiar interpretation.

[1] G has type F_n if there is a $K(G,1)$-complex with finite n-skeleton. All groups have type F_0, F_1 is "finitely generated", F_2 is "finitely presented", etc.

- Let G have type F_n and let the action ρ have discrete orbits. Then ρ is CC^{n-1} if and only if the stabilizers of points of M have type F_n. See Theorem A.

When we come down to specific examples we find new information about some very old group actions. If m stands for a natural number and $s(m)$ for the number of different prime divisors of m we find

- The natural action of the group $G = SL_2(\mathbb{Z}[\frac{1}{m}])$ on the hyperbolic plane $\mathbb{H}^2$ by Möbius transformations is $CC^{s(m)-2}$ but not $CC^{s(m)-1}$.

The lowest case of this, $s(1) = 0$, is the well-known fact that $SL_2(\mathbb{Z})$ does not act cocompactly.

1.3. The Boundary Criterion

The art of deciding whether a given group action $\rho : G \to \mathrm{Isom}(M)$ is CC^n is still in its infancy. An approach via the compactifying boundary ∂M of the CAT(0)-space M is presented in Part II. There we define the analogous concept of ρ being "controlled n-connected (or CC^n) over an endpoint $e \in \partial M$". In rough terms this means (referring back to $h : X \to M$) that the preimages of the horoballs of M at e are coarsely n-connected subsets of X. The precise relationship between the various CC^n-properties is exhibited in the following result

- Let M be almost geodesically complete and let G be of type F_n. Then an isometric action ρ of G on M is CC^{n-1} if and only if ρ is CC^{n-1} over each boundary point $e \in \partial M$.

This "Boundary Criterion" should be viewed in the tradition of local-global principles: It breaks up the problem of deciding whether ρ is CC^{n-1} into "local" questions over each $e \in \partial M$, so that different methods and viewpoints may be used for different kinds of endpoints. In our paper [**BG**] we apply it to establish the above SL_2-theorem; we do not know of a direct proof avoiding the Boundary Criterion. In fact, we strengthen that theorem to (see §10.7B):

- The canonical $SL_2(\mathbb{Z}[\frac{1}{m}])$-action on $\mathbb{H}^2$ is CC^∞ over each irrational boundary point $e \in \partial\mathbb{H}^2$ and is $CC^{s(m)-2}$ but not $CC^{s(m)-1}$ over each rational boundary point e.

1.4. The Geometric Invariants

Whether a given isometric action ρ is CC^{n-1} over $e \in \partial M$ depends, in a delicate way, upon the endpoint e. Therefore the subset of ∂M,

$$\Sigma^n(\rho) = \{e \mid \rho \text{ is } CC^{n-1} \text{ over } e\}$$

becomes interesting in its own right and in Part II we investigate its structure. A key tool (among other things to establish the Boundary Criterion) is the subset $\mathring{\Sigma}^n(\rho) \subseteq \Sigma^n(\rho)$,

$$\mathring{\Sigma}^n(\rho) = \{e \mid \mathrm{cl}_{\partial M}(Ge) \subseteq \Sigma^n(\rho)\}.$$

We prove:

- $\mathring{\Sigma}^n(\rho)$ is an open subset of ∂M with respect to the Tits-metric topology on ∂M, but is not, in general, open with respect to the usual (compact) topology on ∂M.

In the special case when M is Euclidean space $\mathbb{E}^m$ and G acts by translations, i.e. $\rho : G \to \mathrm{Transl}(\mathbb{E}^m)$, the boundary ∂M is the sphere S^{m-1} with the trivial G-action. In this situation our results were previously known: the two sets $\overset{\circ}{\Sigma}{}^n(\rho)$ and $\Sigma^n(\rho)$ coincide (as do the two topologies) and have a description in terms of the Geometric Invariant $\Sigma^n(G)$ on which a considerable literature exists, and which was the starting point of our investigation. In [**BGr 84**], [**Me 94**], [**Me 95**], [**Me 96**], [**Me 97**], [**Geh**], [**Ko 96**], [**BeBr 97**], [**Bu**] and [**MMV 98**] $\Sigma^n(G)$ is computed for metabelian groups, Houghton groups, Borel subgroups and other soluble subgroups of Chevalley groups over function fields with finite base fields, and right angled Artin groups, as well as direct products and graph products. Of course $\Sigma^n(G)$ is vacuous if G has finite abelianization. By contrast our new theory produces interesting $\Sigma^n(\rho)$ in cases such as $\mathrm{SL}_2(\mathbb{Z}[\frac{1}{m}])$ where every non-trivial normal subgroup has finite index.

Part 1

Controlled connectivity and openness results

CHAPTER 2

Outline, Main Results and Examples

2.1. Non-positively curved spaces

We interpret "non-positively curved" to mean that (M, d) is a proper CAT(0) space. In detail: (i) it is a non-empty geodesic metric space: this means that an isometric copy of the closed interval $[0, d(a, b)]$ called a *geodesic segment* joins any two points $a, b \in M$; (ii) for any geodesic triangle Δ in M with vertices a, b, c let Δ' denote a triangle in the Euclidean plane with vertices a', b', c' and corresponding side lengths of Δ' and Δ equal; let ω and ω' be geodesic segments from b to c and from b' to c' respectively; then for any $0 \leq t \leq d(b, c)$, $d(a, \omega(t)) \leq ||a' - \omega'(t)||$; and (iii) d is *proper*, i.e. the closed ball $B_r(a)$ around any $a \in M$ of any radius r is compact.

In a CAT(0) space the geodesic segment from a to b is unique and varies continuously with a and b. This implies that CAT(0) spaces are contractible.

For one of our results we only require a weaker property: that the (non-empty) proper metric space have unique geodesic segments – in which case we say M is a *unique-geodesic metric space.* Indeed this is enough to imply that geodesic segments vary continuously with their end points, hence contractibility ([**BrHa**, I(3)]).

Examples of CAT(0)-spaces are Euclidean space $\mathbb{E}^m$, hyperbolic space $\mathbb{H}^m$, locally finite affine buildings, complete simply connected open Riemannian manifolds of non-positive sectional curvature, and any finite cartesian product $(\prod_i M_i, d)$ of CAT(0) spaces (M_i, d_i) with $d(a, b) := (\sum_i d_i(a_i, b_i)^2)^{\frac{1}{2}}$.

2.2. Controlled connectivity: the definition of CC^{n-1}

In this memoir we are given a non-negative integer n, a (discrete) group G of type[1] F_n, and a proper unique-geodesic metric space M – we will eventually be forced to assume that M is CAT(0) but not yet. When $\rho : G \to \mathrm{Isom}(M)$ is an action[2] of G on M by isometries, we wish to define the statement "ρ is controlled $(n-1)$-connected". To do this we pick an n-dimensional $(n-1)$-connected CW complex X^n on which G acts freely as a group of cell permuting homeomorphisms with $G \backslash X^n$ a finite complex. In other words X^n is the n-skeleton of a contractible free G-CW complex X which is chosen so that X^n is cocompact. Then we choose a G-equivariant continuous function $h : X \to M$. We call such an equivariant h a *control function.* Control functions exist because X is G-free and M is contractible. For $a \in M$ and $r \geq 0$ we denote by $X_{(a,r)}$ the largest subcomplex of X lying in $h^{-1}(B_r(a))$. We say X is *controlled* $(n-1)$*-connected* (CC^{n-1}) over M with respect

[1]It is possible to define CC^n for actions of G on M by isometries even when G is not of type F_n; see Appendix B.

[2]In this memoir all group actions are left actions unless otherwise stated.

to h if for every $r \geq 0$ and $p \leq n-1$ there exists $\lambda \geq 0$ such that for every $a \in M$ every map $f : S^p \to X^n_{(a,r)}$ extends to a map $\tilde{f} : B^{p+1} \to X^n_{(a,r+\lambda)}$. The number λ is independent of a but depends on r and p; however, its dependence on p can be removed trivially since there are only finitely many relevant p. We will sometimes write $\lambda(r)$ for emphasis. We call λ a *lag*. We prove in Chapter 3 that this definition is independent of both X and h: in other words it is a property of the action ρ. Anticipating that, we will simply say[3] that ρ is CC^{n-1} if X is CC^{n-1} over M with respect to h.

EXAMPLE 2.1. Let $G = F_2$ the free group of rank 2. The abelianization homomorphism $\phi : F_2 \twoheadrightarrow \mathbb{Z}^2 \subset \mathbb{R}^2$ defines an action of F_2 on the Euclidean plane by integral translations. Take X to be the Cayley graph of F_2 with respect to a basis $\{a, b\}$. Thus X is an F_2-tree whose vertex set is the free transitive F_2-set F_2. Form the F_2-map $h : X \to \mathbb{R}^2$ by extending ϕ affinely on each edge. For any lag $\lambda \in \mathbb{N}$ there are points in $X_{(0,0)}$ which cannot be joined in $X_{(0,\lambda)}$; for example the vertex 1 and the commutator $[a, b^{\lambda+1}]$. Thus X is not CC^0 over $\mathbb{R}^2$ with respect to h and this action of F_2 on $\mathbb{R}^2$ is not CC^0.

The definition of CC^{n-1} involved all integers $p \leq n-1$. The cases $p \leq -2$ are without content – i.e. are always satisfied, but $p = -1$ is non-trivial. We show in §3.1 that CC^{-1} is equivalent to cocompact,[4] hence whenever ρ is CC^{n-1} ρ is cocompact. The above example is cocompact, so it is CC^{-1} but not CC^0.

EXAMPLE 2.2. The previous example is an action of F_2 on $\mathbb{R}^2$ which is not CC^0. To see actions which are CC^0, take any group G which acts freely, properly discontinuously[5] and cocompactly on a CAT(0) space M where $G \backslash M$ admits the structure of a CW complex. Then taking $X = M$ with the induced CW complex structure, and $h =$ identity, this free G-action on M is CC^0, indeed CC^∞, because balls in CAT(0) spaces are contractible. For example, take $G = F_2$, and $M = X$ as above.

2.3. The case of discrete orbits

Before we state our main results we interpret what CC^{n-1} means when ρ has discrete orbits[6]

THEOREM A. *Let (M, d) be a proper unique-geodesic metric space, let $a \in M$, let G be a group of type F_n, and let $\rho : G \to \mathrm{Isom}(M)$ have discrete orbits. Then ρ is CC^{n-1} if and only if ρ is cocompact and the stabilizer G_a has type F_n.*

It follows that all stabilizers $G_a, a \in M$, have the same F_m-type – but that does not come as a surprise since the point stabilizers of a discrete action by isometries are easily seen to be pairwise commensurable,[7] so one has type F_m if and only if the other has type F_m.

---

[3]Previous publications on the geometric invariants used the terminology "$X_{(a,r)}$ is essentially $(n-1)$-connected" rather than "X is CC^{n-1}". This concept also appears in papers on controlled topology, e.g. [**FePe 95**]. Alternative formulations of this condition are found in the Appendices.

[4]See §1.1 for the definition of "cocompact" and §3.1 for a discussion of CC^{-1}.

[5]An action of Q on M is *properly discontinuous* if every point $a \in M$ has a neighborhood U such that $\{q \in Q | qU \cap U \neq \emptyset\}$ is finite (equivalently: if the action has discrete orbits and has finite point stabilizers, see Lemma 8.5).

[6]An action of G on M *has discrete orbits* if every orbit is a closed discrete subset of M.

[7]$G_{\{a,b\}}$ has finite index in G_a and G_b because orbits are discrete and M is proper.

Theorem A is a consequence of the following homotopy version of K.S. Brown's finiteness criterion [**Br 87$_{\mathbf{I}}$**, Theorem 2.2]:

F_n-CRITERION. *Let H be a group, Y a contractible free H-CW complex and $(K_r)_{r\in\mathbb{R}}$ an increasing filtration of Y by H-subcomplexes so that $Y = \bigcup_r K_r$ and each K_r has cocompact n-skeleton. Then H is of type F_n if and only if Y is CC^{n-1} with respect*[8] *to the filtration (K_r).*

We discuss the proof of this criterion in Chapter 8.

Theorem A follows by setting $Y = X$, $K_r = X_{(a,r)}$ and $H = G_a$. Clearly $X_{(a,r)}$ is a G_a-subcomplex. The remaining part of the proof, that each $X^n_{(a,r)}$ is cocompact as a G_a-complex, is not hard and is given in Chapter 8.

A special case of Theorem A is worth noting. If $N = \ker\rho$ we have a short exact sequence $N \rightarrowtail G \twoheadrightarrow Q$ with $Q \leq \mathrm{Isom}(M)$, and short exact sequences for the stabilizers $N \rightarrowtail G_a \twoheadrightarrow Q_a$. If we replace the assumption that ρ have discrete orbits by the stronger assumption that the induced action of Q on M be properly discontinuous then Theorem A applies – but since all Q_a are finite it follows that each G_a is of type F_n if and only if N is of type F_n. Hence Theorem A becomes

THEOREM A′. *Let Q act properly discontinuous on M. Then ρ is CC^{n-1} if and only if ρ is cocompact and N has type*[9] *F_n.* □

2.4. The Openness Theorem

The main result in this first Part is concerned with general – not necessarily discrete – actions.

THEOREM B. *Let (M, d) be a proper* CAT(0) *space, let n be a non-negative integer and let G be a group of type F_n. The set of all isometric actions of G on M which are CC^{n-1} is an open subset of* $\mathrm{Hom}(G, \mathrm{Isom}(M))$. *In particular, taking $n = 0$, the set of all cocompact actions is open.*

Here, $\mathrm{Isom}(M)$ is the topological group of isometries of M and $\mathrm{Hom}(G, \mathrm{Isom}(M))$ is the space of homomorphisms of the discrete group G into $\mathrm{Isom}(M)$; both function spaces have the compact-open topology. Theorem B is Theorem 7.7 below.

COROLLARY C. *Let $\mathcal{R}(G, M)$ denote the space of all isometric actions of G on M which have discrete orbits. Then the set of all isometric actions $\rho \in \mathcal{R}(G, M)$ which are cocompact and have point stabilizers of type F_n is open in $\mathcal{R}(G, M)$.*

COROLLARY C′. *Let $\mathcal{R}_0(G, M)$ denote the subspace of all $\rho \in \mathcal{R}(G, M)$ with the property that $\rho(G)$ acts properly discontinuously on M. Then the set of all $\rho \in \mathcal{R}_0(G, M)$ which are cocompact and have* $\ker\rho$ *of type F_n is open*[10] *in $\mathcal{R}_0(G, M)$.*

There is no hope of a general openness result in $\mathrm{Hom}(G, \mathrm{Isom}(M))$ for the finiteness properties "$\ker\rho$ is of type F_n" or "the point stabilizers of ρ are of type

[8]In §2.2 we defined CC^{n-1} using filtrations which came from control functions, but the definition makes sense with respect to any filtration.

[9]In the special case when Q is abelian and acts canonically by translations on the vector space $Q \otimes_{\mathbb{Z}} \mathbb{R}$ the combination of Theorem A′ and the Boundary Criterion (Theorem H) is one of the central results of BNSR theory. See Theorem A of [**BS 80**], Theorem B of [**BNS 87**] and Theorem B of [**Re 88**].

[10]Corollary C′ has predecessors in the literature for the case of homomorphisms $\rho : G \to \mathbb{Z}$. Openness of the condition "$\ker\rho$ is finitely generated" was proved in [**Ne 79**], and of the condition "$\ker\rho$ is finitely presented" in [**FrLe 85**]. See also [**BRe 88**] and [**Re 88**]

F_n". This indicates the advantage of the property CC^{n-1} over these traditional finiteness properties. To get a counterexample, consider a finitely generated group G whose abelianization G/G' is free of rank 2, and take M to be the Euclidean line. Then every non-discrete translation action of G on $\mathbb{E}^1$ has kernel the commutator subgroup G'. But the non-discrete translation actions are dense in the space of all translation actions. So if we had an openness result for the property "$\ker \rho$ is finitely generated", it would imply "G' is finitely generated if (and only if) some homomorphism $\chi : G \twoheadrightarrow \mathbb{Z}$ has finitely generated kernel". This is absurd as is shown by the direct product $G = \langle a, x | xax^{-1} = a^2 \rangle \times \mathbb{Z}$ which has commutator subgroup isomorphic to the dyadic rationals, i.e. $G' \cong \mathbb{Z}[\frac{1}{2}]$.

2.5. Connections with Lie groups and local rigidity

The following examples explain how our openness results Theorem B and Corollaries C and C′ are related to locally rigid isometric actions of discrete groups on classical symmetric spaces.

EXAMPLE 2.3. Let M be a locally symmetric space of non-compact type (e.g. the quotient of a virtually connected non-compact linear semisimple Lie group by a maximal compact subgroup). The natural Riemannian metric makes M a proper CAT(0) space. The group $\mathrm{Isom}(M)$ is a Lie group. Call its Lie algebra $\mathfrak{g}$. Each representation $\rho \in \mathrm{Hom}(G, \mathrm{Isom}(M))$ makes $\mathfrak{g}$ into a $\mathbb{Z}G$-module which we denote by $\mathfrak{g}(\rho)$. A theorem of Weil [**We 64**] says that if G is finitely generated and if $H^1(G; \mathfrak{g}(\rho)) = 0$ then all nearby representations are conjugate to ρ in $\mathrm{Isom}(M)$, i.e. ρ has a neighborhood N in $\mathrm{Hom}(G, \mathrm{Isom}(M))$ such that every $\rho' \in N$ is of the form $\rho'(g) = \gamma\rho(g)\gamma^{-1}$ where γ (dependent on ρ') is an isometry of M; then ρ is said to be *locally rigid* (see [**Ra**, p. 90]). In that case $\ker(\rho') = \ker(\rho)$ for all $\rho' \in N$ – a much stronger statement than the conclusion of Corollary C′. But Corollary C′ holds in situations where $H^1(G; \mathfrak{g}(\rho)) \neq 0$, so one may wish to think of it as a weak form of local rigidity: the kernels may not be locally constant, but their finiteness properties are locally constant. The next example illustrates this:

EXAMPLE 2.4. Let G be the group presented by $\langle x, y | xy^2 = y^2x \rangle$. For $n \geq 0$ define $\rho_n : G \to \mathbb{Z}$ by $\rho_0(x) = 0, \rho_0(y) = 1$, and when $n \geq 1$ $\rho_n(x) = n,\ \rho_n(y) = n^2$. It is shown in [**BS**] that $\ker(\rho_0)$ is a free group of rank 2 and when $n \geq 1, \ker(\rho_n)$ is a free group of rank $n^2 + 1$. For $n \geq 1$ define $\tilde{\rho}_n : G \to \mathbb{R}$ by $\tilde{\rho}_n(g) = \frac{1}{n^2}\rho_n(g)$. Identifying $\mathbb{R}$ with the translation subgroup of $\mathrm{Isom}(\mathbb{R})$, we see that $\{\tilde{\rho}_n\}$ converges to ρ_0 in $\mathrm{Hom}(G, \mathrm{Isom}(\mathbb{R}))$, and $\ker(\tilde{\rho}_n) = \ker(\rho_n)$ for all $n \geq 1$. Indeed, each $\tilde{\rho}_n$ is a cocompact action and $\tilde{\rho}_n(G)$ acts properly discontinuously on $\mathbb{R}$. This is a case where Corollary C′ applies but local rigidity fails.

REMARK. The paper [**Fa 99**] contains results in a Lie group context which can be seen as analogous to Theorem B and Corollary C′.

2.6. The new tool

If X and Y are two CW complexes, we write $\hat{\mathcal{F}}(X, Y)$ for the set of all cellular maps $f : D(f) \to Y$, where $D(f)$ is a finite subcomplex of X. By a *sheaf of maps on X with values in Y* we mean[11] any subset $\mathcal{F}$ of $\hat{\mathcal{F}}(X, Y)$ which is closed under restrictions and finite unions. The sheaf $\mathcal{F}$ is *complete* (resp. *locally finite*) if each

[11] A more leisurely exposition of all this is given in Chapter 4. Here, we say just enough about sheaves and finitary maps to state Theorem D.

finite subcomplex of X occurs as the domain of some member (resp. finitely many members) of $\mathcal{F}$. *A cross section* of the complete sheaf $\mathcal{F}$ is a map $X \to Y$ whose restrictions to all finite subcomplexes lie in $\mathcal{F}$. Every cellular map $\phi : X \to Y$ is a cross section of its "restriction", the sheaf $\text{Res}(\phi)$ consisting of all restrictions of ϕ to finite subcomplexes.

These concepts become useful if X and Y are endowed with cell permuting actions of a group G. Then $\hat{\mathcal{F}}(X,Y)$ has a natural G-action: If $g \in G$ and $f \in \hat{\mathcal{F}}(X,Y)$ then the g-translate of f, which we write $gf \in \hat{\mathcal{F}}(X,Y)$, has domain $D(gf) = gD(f)$ and maps gx to $gf(x)$ for each $x \in D(f)$. A *G-sheaf* is a sheaf which is invariant under this action. If $\phi : X \to Y$ is a G-equivariant cellular map then $\text{Res}(\phi)$ is a G-sheaf and is, of course, locally finite. If ϕ is an arbitrary cellular map then the G-sheaf generated by $\text{Res}(\phi)$ will not, in general, be locally finite. But if it is so – and the important fact is that this happens far beyond the equivariant case – we call ϕ a *finitary* (more precisely: *G-finitary*) map. Thus a finitary map $\phi : X \to Y$ is just a cellular map which can be exhibited as a cross section of a locally finite G-sheaf.

In our situation, finitary maps will occur as cellular endomorphisms $\phi : X^n \to X^n$ of the free, n-dimensional, $(n-1)$-connected, cocompact G-CW complex X^n of §2.2. Recall that X^n is endowed with a chosen G-equivariant control map $h : X^n \to M$ into the CAT(0)-space M. The key result in the proof of Theorem B expresses the CC^{n-1} condition of X^n over a in terms of the following "dynamical condition" in X^n. We call a cellular map $\phi : X^n \to X^n$ a *contraction (towards a)* if there exists a radius $r \geq 0$ and a number $\varepsilon > 0$ such that

$$d(a, h\phi(x)) \leq d(a, h(x)) - \varepsilon, \text{ for every } x \in X^n \text{ outside } X^n_{(a,r)}.$$

This is independent of a. We prove

THEOREM D. *Assume the action of G on (M,d) is cocompact. Then X^n is CC^{n-1} if and only if there exists a G-finitary contraction $\phi : X^n \to X^n$.*

This is contained in Theorem 6.8.

2.7. Summary of the core idea

The control function h can be chosen to vary continuously with ρ. Let a given cocompact action ρ of G on M be CC^{n-1}, so that we have a finitary contraction $\phi : X^n \to X^n$ as in Theorem D. If we could describe ϕ in terms of a finite number of equations we might expect that the very same ϕ would still fulfill the properties asserted in Theorem D if the action ρ were subjected to a small perturbation. However, a description of ϕ requires not only the finitary G-sheaf $\mathcal{F}(\phi)$ generated by $\text{Res}(\phi)$ but also an infinite number of choices of members of $\mathcal{F}(\phi)$. Thus we cannot expect the same ϕ to work for all ρ' near ρ. But the sheaf $\mathcal{F}(\phi)$ itself can be described in terms of a finite number of equations and we can pin down a finite number of inequalities which are necessary and sufficient for $\mathcal{F}(\phi)$ to have a cross section which fulfills Theorem D. Thus, even though perturbing the action ρ slightly to ρ' requires a new finitary contraction ϕ' to establish CC^{n-1} for ρ', we will be able to guarantee that ϕ' does exist as a cross section of the old sheaf $\mathcal{F}(\phi)$.

2.8. SL_2 examples

Let K be a field endowed with a discrete valuation $v : K \to \mathbb{Z} \cup \{\infty\}$. Then we can take M to be the Bruhat-Tits-tree of $\text{SL}_2(K)$ (see [**Se**], Chapitre II) acted on by $\mathrm{GL}_2(K)$. Every representation $\rho : G \to \mathrm{GL}_2(K)$ can thus be interpreted as an action of G on M with discrete orbits. Let S be a finite set of pairwise inequivalent discrete valuations containing v. Let $O_S \subseteq K$ denote the ring of S-integers,[12] put $G = \mathrm{SL}_2(O_S)$ and take ρ_0 to be the natural embedding of G into $\mathrm{GL}_2(K)$. The action ρ_0 is cocompact as long as $|S| \geq 2$.

In the case when K is a finite extension of the field of rational functions $\mathbb{F}_q(t)$ over a finite field $\mathbb{F}_q$ we know by a result of U. Stuhler [**St 80**] that G is of type $F_{|S|-1}$ but not of type $F_{|S|}$. This applies also to the point stabilizers of ρ_0 which are $\mathrm{SL}_2(O_S) \cap \mathrm{GL}_2(O_v) = \mathrm{SL}_2(O_{S-\{v\}})$, and hence the point stabilizers of ρ_0 are of type $F_{|S|-2}$ but not $F_{|S|-1}$. By Theorem A we conclude that ρ_0 is $CC^{|S|-3}$ but not $CC^{|S|-2}$.

The interesting point here is that a similar phenomenon seems to occur in the parallel case when S is a finite set of rational primes and $G = \mathrm{SL}_2(\mathbb{Z}_S)$ acts by Moebius transformations on the hyperbolic plane $M = \mathbb{H}^2$. This situation is of course more subtle since this action does not have discrete orbits when $|S| \geq 1$ so that Theorem A is not applicable. Nevertheless we have proved[13] that the natural action of $G = \mathrm{SL}_2(\mathbb{Z}_S)$ on $M = \mathbb{H}^2$ is $CC^{|S|-2}$ but not $CC^{|S|-1}$. The proof requires consideration of the CC^{n-1}-property over endpoints of M. That is the theme of Part II; this discussion of $\mathrm{SL}_2(\mathbb{Z}_S)$ is continued in §10.7(B).

The rest of Part I consists of proofs of what has been outlined here. An outline of Part II is found in Chapter 10.

---

[12] O_S consists of all $x \in K$ with $w(x) \geq 0$ for all valuations w on K except possibly those in S.

[13] One expects the finiteness properties of S-arithmetic groups to be quite different over function fields than over number fields. For instance, "F_{k-1} but not F_k" does not occur for reductive groups in the number field case whereas it is typical in the function field case. Our result indicates that "CC^{k-1} but not CC^k" appears in the number field case, making the two cases more similar.

CHAPTER 3

Technicalities Concerning the CC^{n-1} Property

3.1. Local and global versions of CC^{n-1}

Let (M, d) be a proper unique-geodesic space. In keeping with §2.2, X is an $(n-1)$-connected free G-complex of dimension $\geq n$ with $G\backslash X^n$ finite, and ρ is an action of G on M by isometries.

PROPOSITION 3.1. *There exists a control function (i.e. G-map $h : X \to M$). If h_1 and h_2 are two such then* $\sup\{d(h_1(x), h_2(x)) | x \in X^n\} < \infty$.

PROOF. The map h is defined inductively on skeleta. On X^0 define h arbitrarily on one representative vertex in each orbit of vertices, and then extend equivariantly. Assuming h defined on X^{k-1}, choose a representative k-cell σ in each orbit of k-cells. Since M is contractible, for each such σ the map $h|\overset{\bullet}{\sigma}$ can be extended[1] to σ. Then h can be extended equivariantly to the rest of X^k. The second part is clear since $G\backslash X^n$ is finite. □

Choose a control function $h : X \to M$. The property that X is CC^{n-1} over M with respect to h was defined in §2.2. Given $a \in M$ we say that *X is CC^{n-1} over a with (respect to h)* if for every $r \geq 0$ and $p \leq n-1$ there exists $\lambda \geq 0$ such that every map $f : S^p \to X^n_{(a,r)}$ extends to a map $\tilde{f} : B^{p+1} \to X^n_{(a,r+\lambda)}$. The difference is that in this definition $\lambda = \lambda(a, r)$ (dependence on p being removed trivially as before). It is an immediate consequence of the triangle inequality in M that if X is CC^{n-1} over some $a \in M$ with lag λ then X is CC^{n-1} over any $b \in M$ with lag $\lambda + 2d(a, b)$. Hence the property "CC^{n-1} over a" is actually independent of the choice of $a \in M$ even though we have to expect that the lag λ does depend on a as well as on the radius r.

We are going to show that when ρ is a cocompact action, CC^{n-1} over M is equivalent to CC^{n-1} over one point $a \in M$. In preparation, it will be helpful to say what our definitions mean when $n = 0$. "(-1)-connected" means "non-empty": the sphere S^{-1} is empty and has a unique empty map $S^{-1} \to X$, which extends to the ball B^0 if and only if X is non-empty. To say that X is CC^{-1} over a is simply to say that X^0 is non-empty (something which has already been assumed since dim $X \geq n \geq 0$)[2].

PROPOSITION 3.2. (i) *X is CC^{-1} over M if and only if ρ is cocompact;* (ii) *X is CC^{n-1} over M if and only if ρ is cocompact and X is CC^{n-1} over some $a \in M$.*

PROOF. If X is CC^{-1} over M then there exists $\lambda \geq 0$ such that for every $a \in M$ $h(X^0) \cap B_\lambda(a) \neq \emptyset$. Since $G\backslash X^0$ is finite and non-empty we may choose representatives $v_1, \ldots, v_m$ of the orbits in X^0. Let $K_0 = \{h(v_1), \ldots, h(v_m)\}$. Then

[1] $\overset{\bullet}{\sigma}$ denotes the cell-boundary of σ, i.e., $\overset{\bullet}{\sigma} = \sigma \cap X^{k-1}$

[2] Note that we have tacitly included the value $p = -1$ here and in §2.2.

$GK_0 = h(X^0)$. If K is the closed λ-neighborhood of K_0 then K is compact and the opening sentence can be restated as $GK = M$. Hence ρ is cocompact. Conversely, if $M = GK$ for some compact subset K of M, let λ be the diameter of K. We may assume $h(v) \in K$ for some $v \in X^0$. Given $a \in M$ there is some $g \in G$ such that $a \in gK$. Hence $d(a, gv) \leq \lambda$ so $X^0_{(a,\lambda)} \neq \emptyset$ with λ independent of a. Thus X is CC^{-1} over M. This proves (i).

To prove (ii) assume X is CC^{n-1} over $a \in M$ with lag $\lambda = \lambda(a,r)$ and ρ is cocompact. We have $X_{(ga,r)} = gX_{(a,r)}$, for $g \in G$ and $r \geq 0$, so λ is a lag for each $a' \in Ga$. Hence $\lambda + \min 2d(Ga, b)$ is a lag for $b \in M$, and if ρ is cocompact this has an upper bound independent of b. The other direction of (ii) is trivial in view of (i), so (ii) is proved. □

3.2. The Invariance Theorem

Up to now we have defined the property CC^{n-1} of actions of G on M by isometries using an n-dimensional $(n-1)$-connected cocompact free G-CW-complex X^n and a control function $h : X^n \to M$. We must prove invariance: that the property is independent of the choices of X and h. In some cases a natural control function presents itself on a non-free G-CW-complex.[3] It is useful to be able to read off the CC^{n-1} property directly in such a case:

THEOREM 3.3. *Let G be of type F_n, let ρ be an action of G on M by isometries, and let Y be a cocompact n-dimensional $(n-1)$-connected rigid G-CW-complex such that the stabilizer of each p-cell is of type F_{n-p}, $p \leq n-1$. Let $h : Y \to M$ be a G-map and let $a \in M$. The property that Y be CC^{n-1} over a is independent of the choices of a, of Y and of the G-map h.*

By 3.2 it follows that "ρ is CC^{n-1}" is independent of Y and of h.

In case Y is a free G-complex, Theorem 3.3 has an elementary proof which we sketch below. When Y is not free a more difficult proof is required which we delay until Chapter 9 because the methods are not related to anything else in this memoir.

PROOF OF THEOREM 3.3 WHEN Y IS FREE (SKETCH). Independence of a is clear. Independence of h follows from Proposition 3.1. In the free case we may always attach cells to make the complex contractible, and extend h. Let X and Y be two contractible free G-CW complexes with cocompact n-skeleta. Choose cellular maps $\alpha : G\backslash X \to G\backslash Y$ and $\beta : G\backslash Y \to G\backslash X$ which are mutually homotopy inverse. Their lifts $\tilde{\alpha} : X \to Y$ and $\tilde{\beta} : Y \to X$ are bounded maps on the n-skeleta and there is a bounded homotopy[4] in X^n between $\tilde{\alpha} \circ \tilde{\beta}|Y^{n-1}$ and the inclusion map $Y^{n-1} \hookrightarrow Y^n$. If $h : X \to M$ is a G-map then $h \circ \tilde{\beta} : Y \to M$ is a G-map. It follows directly that if X is CC^{n-1} over a (using h) then Y is CC^{n-1} over a (using $h \circ \tilde{\beta}$). □

As an application of Theorem 3.3 we find

[3] By a *G-CW-complex* we mean a CW complex Y equipped with a G-action which permutes cells. If this action has the additional property that for any point x in the interior of any cell σ the stabilizers G_x and G_σ are equal, we call Y a *rigid* G-CW-complex.

[4] A map [resp. homotopy] is *bounded* if there is an integer N such that for all cells σ of $\tilde{X}^{n-1}$, the image of σ [resp. of $\sigma \times I$] lies in a subcomplex containing $\leq N$ cells.

COROLLARY 3.4. *Let $H \leq G$ be a subgroup of finite index in G. Then the G-action ρ is CC^{n-1} if and only if its restriction $\rho \mid H$ is CC^{n-1}.*

PROOF. The G-CW-complex X and the control function $h : X \to M$ can also be used to test the CC^{n-1} property of the restricted action $\rho \mid H$. Since the subcomplexes $X_{(a,r)}$ remain the same so do the CC^{n-1} properties. $\square$

CHAPTER 4

Finitary Maps and Sheaves of Maps

This is a self-contained introduction to a new topological tool. The important idea—finitary maps[1]—is introduced in §4.7. A finitary map is a special kind of map between G-CW complexes which generalizes the notion of equivariant map. In our situation there are not enough equivariant maps but there are enough finitary maps.

4.1. Sheaves of maps

Let X and Y be CW-complexes. By a *sheaf (of maps)* $\mathcal{F} : X \rightsquigarrow Y$ we mean a set $\mathcal{F}$ of cellular maps $f : D(f) \to Y$ with domain $D(f)$ a finite subcomplex of X satisfying the following axioms:

(i) $\mathcal{F}$ contains the empty map.
(ii) If $f \in \mathcal{F}$ and if K is a subcomplex of $D(f)$ then $f \mid K$ is also in $\mathcal{F}$.
(iii) If f and f' are in $\mathcal{F}$ and agree on the intersection of their domains then $f \cup f' : D(f) \cup D(f') \to Y$ is also in $\mathcal{F}$.

If K is a subcomplex of X, $\mathcal{F}|K$ denotes the sheaf consisting of all restrictions of maps in $\mathcal{F}$ to subcomplexes of K. A *subsheaf* of a sheaf $\mathcal{F}$ is a subset which is itself a sheaf. Every set of maps from finite subcomplexes of X to Y generates a sheaf in the obvious way. Each sheaf $\mathcal{F} : X \rightsquigarrow Y$ has natural "minimal" generators, namely the members $f \in \mathcal{F}$ whose domains $D(f)$ are carriers[2] of cells of X.

An important example is the sheaf of a cellular map $\phi : X \to Y$ which we denote by $\operatorname{Res}(\phi)$; it consists of all restrictions of ϕ to the finite subcomplexes of X. More generally, if Φ is an arbitrary family of cellular maps $\phi : D(\phi) \to Y$, where each $D(\varphi)$ is a subcomplex of X, we write $\operatorname{Res}(\Phi)$ for the sheaf generated by the restrictions of the members of Φ to finite subcomplexes of X.

By a *cross section of a sheaf* $\mathcal{F}$ we mean a cellular map $\phi : X \to Y$ whose sheaf $\operatorname{Res}(\phi)$ is a subsheaf of $\mathcal{F}$. Cross sections are easy to construct when the sheaf $\mathcal{F}$ is *homotopically closed*; by this we mean that for each $f \in \mathcal{F}$ and each finite subcomplex $K \supseteq D(f)$ there is some $\tilde{f} \in \mathcal{F}$ with $D(\tilde{f}) = K$ and $\tilde{f} \mid D(f) = f$. An easy induction on the skeleta of K shows that for $\mathcal{F}$ to be homotopically closed it suffices that for each n-cell σ of X and each $f \in \mathcal{F}$ with $D(f) = C(\overset{\bullet}{\sigma})$ there is some $\tilde{f} \in \mathcal{F}$ with $D(\tilde{f}) = C(\sigma)$ extending f. And with the same inductive argument one shows that every homotopically closed sheaf $\mathcal{F} : X \rightsquigarrow Y$ admits a cross section $\phi : X \to Y$.

[1]Seminal versions of the concepts of finitary maps, sheaves and homotopies are contained in [**BS 80**], [**BNS 87**] and [**Re 88**], specifically in the proofs of results relating the geometric invariants to finiteness properties of normal subgroups $N \lhd G$ with G/N abelian.

[2]If $A \subset X$, $C(A)$ denotes the smallest subcomplex of X containing A; it is called the *carrier* of A. When A is compact, $C(A)$ is a finite subcomplex. When σ is a cell of X, $C(\sigma)$ has only the cell σ in the top dimension. Clearly $C(\sigma) = C(\overset{\bullet}{\sigma}) \cup \sigma$ and $\overset{\bullet}{\sigma} = C(\overset{\bullet}{\sigma}) \cap \sigma$, so $\dim C(\overset{\bullet}{\sigma}) < \dim C(\sigma)$.

4.2. G-sheaves

Assume that X and Y are G-CW-complexes. Then G acts on the set of all cellular maps $\phi : K \to Y$ with K a subcomplex of X: if $g \in G$ we write $g\phi$ for the g-translate of ϕ; it has domain $D(g\phi) = gK$ and maps $x \in gK$ to $g\phi(g^{-1}x)$. By a *G-sheaf* we mean a sheaf $\mathcal{F} : X \rightsquigarrow Y$ which is invariant under this action. If $\mathcal{F}$ is a sheaf we write $G\mathcal{F}$ for the G-sheaf generated by $\mathcal{F}$. Thus $G\mathcal{F}$ is the set of all maps which can be written as the union of finitely many maps $g_i f_i$, $g_i \in G$, $f_i \in \mathcal{F}$.

By a *fundamental domain* of the G-CW-complex X we mean any subcomplex $F \subseteq X$ with $GF = X$. A sheaf $\mathcal{F}_0 : F \rightsquigarrow Y$ defined on a fundamental domain is *G-saturated* if whenever $f \in \mathcal{F}_0$ and $g \in G$ are such that $g\, D(f) \subseteq F$ then $gf \in \mathcal{F}_0$. The restriction $\mathcal{F}|F$ of any G-sheaf $\mathcal{F} : X \rightsquigarrow Y$ is certainly G-saturated. Conversely, G-saturated sheaves lead to G sheaves as follows:

PROPOSITION 4.1. *Let $F \subseteq X$ be a fundamental domain and $\mathcal{F}_0 : F \rightsquigarrow X$ a sheaf. Then $\mathcal{F}_0$ is G-saturated if and only if $\mathcal{F}_0 = G\mathcal{F}_0|F$.*

PROOF. It is clear that $\mathcal{F}_0$ is always a subsheaf of the G-saturated sheaf $G\mathcal{F}_0|F$. If $f \in G\mathcal{F}_0|F$ then $f = g_1 f_1 \cup \ldots \cup g_n f_n$ with $g_i \in G$, $f_i \in \mathcal{F}_0$. Thus $g_i D(f_i) = D(g_i f) \subseteq D(f) \subseteq F$; but if $\mathcal{F}_0$ is G-saturated this implies $g_i f_i \in \mathcal{F}_0$ and hence $f \in \mathcal{F}_0$. □

Here is a useful fact about sheaves generated by cellular maps:

PROPOSITION 4.2. *Let Φ denote a set of cellular maps $\phi : D(\phi) \to Y$, $D(\phi) \subseteq X$. Then we have* $\mathrm{Res}(G\Phi) = G\ \mathrm{Res}(\Phi)$. *In particular, if Φ is G-invariant, then* $\mathrm{Res}(\Phi)$ *is a G-sheaf. Moreover, in that case* $\mathrm{Res}(\Phi) = G\ \mathrm{Res}(\Phi|F)$ *for every fundamental domain $F \subseteq X$.*

PROOF. $\mathrm{Res}(G\Phi)$ is generated by all restrictions $(g\phi)|K$ with $g \in G$, $\phi \in \Phi$, and K a finite subcomplex of X. But $(g\phi)|K = g(\phi|g^{-1}K)$, and the right hand side of this equation exhibits generators of $G\ \mathrm{Res}(\Phi)$. This proves the first assertion. For the second assertion note that $\mathrm{Res}(\Phi)$ is generated by the restrictions of the maps $\phi \in \Phi$ to the carriers $C(\sigma)$ of the cells σ of X; and a fundamental domain F will always contain G-translates of these carriers. This shows that if $G\Phi = \Phi$ then $\mathrm{Res}(\Phi) \subseteq G\ \mathrm{Res}(\Phi|F)$. The other inclusion is obvious. □

4.3. Locally finite sheaves

A sheaf $\mathcal{F} : X \rightsquigarrow Y$ is said to be *locally finite* if the restriction of $\mathcal{F}$ to each finite subcomplex K of X is a finite set of maps $\mathcal{F}|K : K \rightsquigarrow Y$. Note that it suffices to assume that the restriction of $\mathcal{F}$ to the carrier of each cell is finite.

PROPOSITION 4.3. *Let $\mathcal{F} : X \rightsquigarrow Y$ be a G-sheaf. If the restricted sheaf $(\mathcal{F}|F) : F \rightsquigarrow Y$ is locally finite for some fundamental domain $F \subseteq X$ then $\mathcal{F}$ is also locally finite.*

PROOF. Let σ be a cell of X, and $g \in G$ with $g\sigma \subset F$. Since F is a subcomplex it contains, in fact the whole of the carrier $C(g\sigma) = gC(\sigma)$, and so $g(\mathcal{F}|C(\sigma)) = \mathcal{F}|C(g\sigma)$ is a subsheaf of $(\mathcal{F}|F)$. Since $\mathcal{F}|F$ is locally finite this subsheaf is finite, hence so is $\mathcal{F}|C(\sigma)$. □

A G-sheaf $\mathcal{F}$ is *finitely generated* if it is generated by the G-translates of a finite subset of $\mathcal{F}$.

Using Proposition 4.1 one easily proves

PROPOSITION 4.4. *Let X and Y be G-CW complexes and let $\mathcal{F} : X \rightsquigarrow Y$ be a G-sheaf.*

a) *If $\mathcal{F}$ is locally finite and the G-action on X is cocompact then $\mathcal{F}$ is finitely generated.*
b) *If $\mathcal{F}$ is finitely generated and each cell of X has finite stabilizer then $\mathcal{F}$ is locally finite.*

□

4.4. Embedding sheaves into homotopically closed sheaves

In order to construct cross sections it will often be important to embed a given sheaf in a homotopically closed sheaf. Now, it is clear that if the complex Y is contractible then every sheaf $\mathcal{F} : X \rightsquigarrow Y$ can be embedded in a homotopically closed sheaf $\tilde{\mathcal{F}} : X \rightsquigarrow Y$, and if $\mathcal{F}$ is a G-sheaf we can choose $\tilde{\mathcal{F}}$ to be a G-sheaf. Given that $\mathcal{F}$ is locally finite it requires care to make $\tilde{\mathcal{F}}$ locally finite. In this paper we shall need the embedding only in the situation when the G-CW-complex X has finite cell stabilizers.

PROPOSITION 4.5. *Assume the G-complex X is locally finite with finite cell stabilizers and Y is contractible. Then every locally finite G-sheaf $\mathcal{F} : X \rightsquigarrow Y$ can be embedded in a homotopically closed locally finite G-sheaf $\tilde{\mathcal{F}} : X \rightsquigarrow Y$.*

PROOF. By Lemma 4.6 below we can choose a fundamental domain $F \subseteq X$ which contains only finitely many members of each G-orbit of cells. Let $\mathcal{F}_0 = (\mathcal{F}|F) : F \rightsquigarrow Y$ be the restricted sheaf. $\mathcal{F}_0$ is locally finite and G-saturated (see Proposition 4.1). If we can embed $\mathcal{F}_0$ in a locally finite G-saturated homotopically closed sheaf $\tilde{\mathcal{F}}_0$ then $\tilde{\mathcal{F}} = G\tilde{\mathcal{F}}_0$ will solve our embedding problem. Indeed, by Proposition 4.1 $\tilde{\mathcal{F}}|F = \tilde{\mathcal{F}}_0$, hence Proposition 4.3 applies to show that $\tilde{\mathcal{F}}$ is locally finite; and "homotopically closed" is also a property which is easily seen to be inherited from the restricted sheaf $\tilde{\mathcal{F}}_0 = \tilde{\mathcal{F}}|F$.

It remains to show that the locally finite and G-saturated sheaf $\mathcal{F}_0 : F \rightsquigarrow Y$ can be embedded in a homotopically closed locally finite G-saturated sheaf $\tilde{\mathcal{F}}_0 : F \rightsquigarrow Y$. We construct $\tilde{\mathcal{F}}_0$ inductively on the p-skeleton of F. The induction starts with $p = -1$ where the empty map will do. So assume $p \geq 0$, and that a homotopically closed locally finite and G-saturated sheaf $\tilde{\mathcal{F}}_0^{p-1} : F^{p-1} \rightsquigarrow Y$ has been constructed which contains $\mathcal{F}_0^{p-1} = \mathcal{F}_0|F^{p-1}$. For each p-cell σ of F and each $f \in \tilde{\mathcal{F}}_0^{p-1}$ with domain $D(f)$ the $(p-1)$-skeleton of the carrier of σ, we choose a new map $\tilde{f} : C(\sigma) \to Y$ extending f – this is possible since Y is p-connected. Adjoining these maps $\tilde{f}$ to $\tilde{\mathcal{F}}_0^{p-1}$, together with all other members $f' \in \mathcal{F}^p$ with $D(f') = C(\sigma)$, yields a sheaf $\mathcal{G} : F^p \rightsquigarrow Y$ which is locally finite, homotopically closed and restricts to $\mathcal{G}|F^{p-1} = \tilde{\mathcal{F}}_0^{p-1}$.

$\mathcal{G}$ is not yet G-saturated. In order to make it G-saturated we have to adjoin, for each p-cell σ of F, the maps gf', where $f' \in \mathcal{G}|C(\sigma)$ and g is an element of G with $g\sigma \subseteq F$. By our careful choice of F there are only finitely many translated cells $g\sigma$ in F, and, since the cell stabilizers of X are finite, only finitely many elements $g \in G$ are needed for a given cell σ. This makes the resulting sheaf $\tilde{\mathcal{F}}_0^p$ locally finite and G-saturated. Since $\tilde{\mathcal{F}}_0^p|F^{p-1} = \mathcal{G}|F^{p-1}$ it is still homotopically closed. □

It remains to supply the proof of

LEMMA 4.6. *Every locally finite G-CW-complex X contains a fundamental domain F with the special feature that F contains only finitely many members of each G-orbit of cells.*

PROOF. Let T be a system of representatives for each G-orbit of cells, and let $F \subseteq X$ be the union of all subcomplexes $C(\sigma)$ with σ running through T. Let τ be a cell in X, and $g \in G$ with $g\tau \in F$. Then there is some $\sigma_g \in T$ with $g\tau \subseteq C(\sigma_g)$, i.e., $\tau \subseteq g^{-1}C(\sigma_g) = C(g^{-1}\sigma_g)$. But τ can only be contained in finitely many subcomplexes of the form $C(g^{-1}\sigma_g)$. Hence there are finitely many cells $\sigma_1, \ldots, \sigma_m \in T$ and finitely many elements $g_1, \ldots, g_m$ with the property that for each $g \in G$ with $g\tau \in F$, there is $j \in \{1, \ldots, m\}$ such that $g^{-1}\sigma_g = g_i^{-1}\sigma_i$ – in other words, $\sigma_g = \sigma_i$ and $gg_i^{-1} \in \mathrm{Stab}(\sigma_i)$. It follows that $g\tau \subseteq C(\sigma_g) = C(\sigma_i)$ for some i; hence there are only finitely many possibilities for $g\tau$ to be in F. □

4.5. Composing sheaves

In Part II we will compose sheaves $\mathcal{F} : X \rightsquigarrow Y, \mathcal{F}' : Y \rightsquigarrow Z$. Now, the set $\mathcal{P}$ consisting of all compositions $f' \circ f$ with $f \in \mathcal{F},\ f' \in \mathcal{F}'$ and $f(D(f)) \subseteq D(f')$ is not, in general, a sheaf – though it does have properties (i) and (ii) of the definition of a sheaf in §4.1. We define $\mathcal{F}' \circ \mathcal{F}$ to be the sheaf generated by $\mathcal{P}$. Thus $\mathcal{F}' \circ \mathcal{F}$ consist of all maps $f'' : D(f'') \to Z$ which can be written as a union $f'' = f_1' f_1 \cup \ldots \cup f_k' f_k$ with $f_i \in \mathcal{F}, f_i' \in \mathcal{F}'$ and $\mathrm{image} f_i \subseteq D(f_i')$ for all i.

PROPOSITION 4.7. (a) *Let $\mathcal{F} : X \to Y$ and $\mathcal{G} : Y \to Z$ be sheaves. If $\mathcal{F}$ and $\mathcal{G}$ are G-sheaves, so is $\mathcal{G} \circ \mathcal{F}$. If $\mathcal{F}$ and $\mathcal{G}$ are locally finite, so is $\mathcal{G} \circ \mathcal{F}$.*
(b) *If $\phi : D(\phi) \to Y$ and $\psi : D(\psi) \to Z$ are cellular maps defined on subcomplexes $D(\phi) \subseteq X, D(\psi) \subseteq Y$ with $\phi(D(\phi)) \subseteq D(\psi)$. Then $\mathrm{Res}(\psi \circ \phi) = \mathrm{Res}(\psi) \circ \mathrm{Res}(\phi)$.*

Both assertions are straightforward. □

4.6. Homotopy of sheaves

Let X and Y be CW-complexes. We consider the unit interval I and the product $X \times I$ with their canonical CW-structures. A *homotopy of sheaves*, $\mathcal{H} : X \times I \rightsquigarrow Y$ is a set of cellular maps $H : D(H) \to Y$, where $D(H) = D_H \times I$ with D_H a finite subcomplex of X, such that $\mathcal{H}$ contains the empty maps and is closed under union and restriction (to subcomplexes of the form $D \times I$). More precisely, we will say that such a set of maps $\mathcal{H}$ is a *homotopy between the sheaves* $\mathcal{F}_0 = \mathcal{H}|X \times \{0\}$ and $\mathcal{F}_1 = \mathcal{H}|X \times \{1\}$, where $\mathcal{F}_0$ and $\mathcal{F}_1$ are regarded as sheaves $X \rightsquigarrow Y$.

Although a homotopy of sheaves $\mathcal{H}$ is not a sheaf in the technical sense (restrictions to subcomplexes other than those of the form $D \times I$ are not considered) we can still use constructions like unions, restrictions and compositions, and we can talk about G-homotopies in the obvious sense.

An obvious necessary condition for the existence of a homotopy between two sheaves $\mathcal{F}_0, \mathcal{F}_1 : X \rightsquigarrow Y$ is that the collections of domains of $\mathcal{F}_0$ and of $\mathcal{F}_1$ coincide. If they do, then contractibility of Y is sufficient for the existence of a homotopy $\mathcal{H} : \mathcal{F}_0 \simeq \mathcal{F}_1$. Parallel to Proposition 4.5 we find

PROPOSITION 4.8. *Assume the G-complex X is locally finite with finite cell stabilizers and Y is contractible. Then any two locally finite G-sheaves $\mathcal{F}_0, \mathcal{F}_1 : X \rightsquigarrow Y$ with the same collections of domains are homotopic via a locally finite G-homotopy $\mathcal{H} : X \times I \rightsquigarrow Y$.*

4.7. Finitary maps

A cellular map $\phi : X \to Y$ between two G-CW-complexes is said to be *finitary* (or, more precisely, *G-finitary*) if ϕ is a cross section of a locally finite G-sheaf $\mathcal{F} : X \rightsquigarrow Y$. Such a G-sheaf $\mathcal{F}$ will have to contain all of $G\text{Res}(\phi)$; hence $G\text{Res}(\phi) = \text{Res}(G\phi)$ is locally finite. This shows that $\phi : X \to Y$ is finitary if and only if the G-sheaf $G\text{Res}(\phi) = \text{Res}(G\phi)$ is locally finite. Hence, in view of Proposition 4.7 we obtain

PROPOSITION 4.9. *If $\phi : X \to Y$ and $\psi : Y \to Z$ are G-finitary maps, so is their composition $\psi \circ \phi : X \to Z$.*

Examples of finitary maps are, of course, the G-equivariant maps $\phi : X \to Y$; for if ϕ is equivariant then $\text{Res}(\phi)$ is a G-sheaf with cross section ϕ. But there are many finitary maps beyond the equivariant ones. Take, for instance, X and Y to be the 0-dimensional free G-space G. A finitary map $\phi : G \to G$ is then given by a finite set of elements $T \subseteq G$ and a function $\kappa : G \to T$; $\phi(g) = g\kappa(g)$, $g \in G$. If one wishes to extend this map ϕ to a finitary map $\phi : \Gamma \to \Gamma$ on the Cayley graph of G with respect to a set of generators $S \to G$, one has to choose a finite set of edge paths P with origin and terminus in T and a map $\mu : G \times S \to P$ such that $\mu(g,s)$ has origin $\kappa(g)$ and terminus $\kappa(gs)$, and put $\phi((g,s)) := \mu(g,s)$.

PROPOSITION 4.10. *If the G-CW-complex X is locally finite with finite cell stabilizers and Y is contractible then any two G-finitary maps ϕ_0, $\phi_1 : X \to Y$ are homotopic via a G-finitary homotopy.*

PROOF. Take locally finite sheaves $\mathcal{F}_0, \mathcal{F}_1 : X \rightsquigarrow Y$ with cross sections ϕ_0, ϕ_1, respectively. By Proposition 4.8 there is a locally finite homotopy $\mathcal{H} : X \times I \rightsquigarrow Y$ between $\mathcal{F}_0$ and $\mathcal{F}_1$; $\mathcal{H}$ is a G-homotopy. By an inductive construction one finds a homotopy $\psi : X \times I \to Y$ from ϕ_0 to ϕ_1 which is a cross section of $\mathcal{H}$. The details are left to the reader. □

CHAPTER 5

Sheaves and Finitary Maps Over a Control Space

In this section we consider sheaves $\mathcal{F} : X \rightsquigarrow X$ of maps on a CW-complex into itself over a "control space" M. Throughout the section M is a proper metric space – the CAT(0) property will not be used. The group G is assumed to act by cell permuting automorphisms on X and by isometries on M, and we are given a control function $h : X \to M$.

5.1. Displacement function and norm

Given a map $f : D(f) \to X$ with $D(f)$ a (not necessarily finite) subcomplex of X we consider

- the *displacement function of f over M*, $\alpha_f : D(f) \to [0, \infty)$ defined by $\alpha_f(x) := d(h(x), hf(x))$, $x \in D(f)$, and
- the *norm of f over M*, $||f|| \in [0, \infty]$, defined by $||f|| := \sup \alpha_f(D(f))$.

The displacement function of f is a continuous non-negative function. As G acts by isometries and h is a G-map it satisfies

$$\alpha_{gf}(gx) = \alpha_f(x), \qquad \text{all} \quad g \in G,\ x \in X. \tag{5.1}$$

The norm $||f||$ may be infinite if $D(f)$ is not finite. If $||f|| < \infty$ we call f *bounded* (over M). We put $||\emptyset|| = 0$ if $\emptyset$ is the empty map.

It is an easy matter to verify that the following formulae hold for all maps $f_i : D(f_i) \to X$ as above and all $g \in G$

$$||gf_i|| = ||f_i|| \tag{5.2}$$

$$||f_1 \cup f_2|| = \max(||f_1||, ||f_2||), \quad \text{provided} \quad f_1 \cup f_2 \quad \text{exists}, \tag{5.3}$$

$$||f_1 \circ f_2|| \leq ||f_1|| + ||f_2||. \tag{5.4}$$

The notion of displacement and norm extend to locally finite sheaves. If $\mathcal{F} : X \rightsquigarrow X$ is a locally finite sheaf and $D(\mathcal{F})$ stands for the union of all domains $D(f)$, $f \in \mathcal{F}$, then $\alpha_{\mathcal{F}} : D(\mathcal{F}) \to [0, \infty)$ is the continuous function $\alpha_{\mathcal{F}}(x) = \sup\{\alpha_f(x) | f \in \mathcal{F}\}$ and $||\mathcal{F}|| := \sup \alpha_{\mathcal{F}}(D(\mathcal{F}))$. Again we call $\mathcal{F}$ *bounded* if $||\mathcal{F}|| < \infty$.

(5.3) and (5.4) show that if $\mathcal{F}$ and $\mathcal{G}$ are locally finite sheaves $X \rightsquigarrow X$ then the norm of their composition satisfies

$$||\mathcal{F} \circ \mathcal{G}|| \leq ||\mathcal{F}|| + ||\mathcal{G}||. \tag{5.5}$$

Useful elementary facts concerning G-sheaves $X \rightsquigarrow X$ and finitary maps $\phi : X \to X$ are collected in

PROPOSITION 5.1. (a) *If $\mathcal{F} : X \rightsquigarrow X$ is a G-sheaf then the displacement function $\alpha_{\mathcal{F}}$ satisfies $\alpha_{\mathcal{F}}(gx) = \alpha_{\mathcal{F}}(x)$ for every $x \in D(\mathcal{F})$.*
(b) *If $G \backslash X$ is compact then every locally finite G-sheaf $\mathcal{F}$ is bounded, and if $\phi : X \to X$ is a cross section of $\mathcal{F}$ then $||\phi|| \leq ||\mathcal{F}||$.*

PROOF. (a) is immediate from (5.1); (5.2) and (5.3) show that the norm of a G-sheaf is given by the norm of a G-generating subset, hence (b). □

5.2. Shift towards a point of M

Let $a \in M$. The *shift* of a map $f : D(f) \to X$, $D(f)$ a subcomplex of X, is defined to be the continuous function $\mathrm{sh}_{f,a} : D(f) \to \mathbb{R}$ given by

$$\mathrm{sh}_{f,a}(x) := d(h(x), a) - d(hf(x), a), \quad x \in D(f).$$

By the triangle inequality we have

$$(5.6) \qquad |\mathrm{sh}_{f,a}(x)\,| \leq \alpha_f(x), \quad \text{all} \quad x \in D(f).$$

Hence the shift function has the global bound $\|f\|$ if f is a bounded map.

If a, b are two points of M then their shift functions are related by the inequality

$$(5.7) \qquad |\mathrm{sh}_{f,a}(x) - \mathrm{sh}_{f,b}(x)\,| \leq 2d(a,b).$$

5.3. Contractions

Let $a \in M$. We call a cellular map $\phi : X \to X$ a *contraction of X towards a* if there are numbers $R \geq 0$, $\varepsilon > 0$ with the property that $\mathrm{sh}_{\phi,a}(x) \geq \varepsilon$ for every $x \in X$ with $d(h(x), a) \geq R$. The number R is an *event radius* for ϕ, and the number ϵ is an *almost guaranteed shift* for ϕ. Using (5.7) one observes that, for each $b \in M$ with $d(a,b) < \frac{\varepsilon}{2}$, ϕ will then also be a contraction towards b (with event radius $R + d(a,b)$ and almost guaranteed shift $\varepsilon - 2d(a,b)$).

PROPOSITION 5.2. *If X admits a contraction towards $a \in M$ with event radius R then X is CC^{-1} over a with lag R.*

PROOF. Let ε be an almost guaranteed shift of this contraction $\phi : X \to X$. Pick any $x \in X^0$ and $m \in \mathbb{N}$ with $m \cdot \varepsilon \geq d(a, h(x)) - R$. Then one of $h(x), h\phi(x), \ldots, h\phi^m(x)$ is in $B_R(a)$, hence the assertion. □

PROPOSITION 5.3. *Let $\phi : X \to X$ be a bounded contraction towards $a \in M$ with event radius R and almost guaranteed shift $\varepsilon > 0$. Then*

$$d(h\phi^m(x), a) \leq \max(d(h(x), a) - m\varepsilon,\ R + \|\phi\|)$$

for each $m \in \mathbb{N}$ and $x \in X$. In particular $\phi^m : X \to X$ is a contraction with event radius $R + \|\phi\| + m\varepsilon$ and almost guaranteed shift $m\varepsilon$.

PROOF. Induction on m using the fact that $d(h\phi(y), a) = d(h(y), a) - \mathrm{sh}_{\phi,a}(y)$ is at most $d(h(y), a) - \varepsilon$, if $d(h(y), a) \geq R$, and at most $R + \|\phi\|$ if $d(h(y), a) \leq R$. □

COROLLARY 5.4. *If a contraction $\phi : X \to X$ towards $a \in M$ exists, then there are contractions $\phi_b : X \to X$ towards each $b \in M$. If G acts cocompactly on M then these ϕ_b can be chosen with uniform event radius R. If ϕ is finitary these ϕ_b can be chosen to be finitary.*

PROPOSITION 5.5. *Assume that the G-CW-complex X is contractible, with finite cell stabilizers and cocompact n-skeleton. If X^n admits a finitary contraction $\phi : X^n \to X^n$ towards a with event radius R and almost guaranteed shift $\varepsilon > 0$ then there is a cellular deformation $\psi : X^n \times [0, \infty) \to X^{n+1}$ and a number $\lambda \geq 0$ such that*

$$d(h\psi(x,s), a) \leq \max\Big(d(h(x), a) - s\varepsilon,\ R + \|\phi\|\Big) + \lambda$$

for every $x \in X^n$ and $s \in [0, \infty)$.

PROOF. By Proposition 4.10 there is a G-finitary homotopy $\psi_0 : X^n \times I \to X^{n+1}$ between Id_{X^n} and ϕ. By analogy with the case of maps, let $||\psi_0|| := \sup_{t\in I} ||\psi_0(\cdot, t)||$. Since ψ_0 is finitary, $||\psi_0|| < \infty$.

Write $s \in [0, \infty)$ as $s = m + t$ where m is a non-negative integer and $t \in [0, 1)$. Define $\psi : X \times [0, \infty) \to X$ to be the (continuous) cellular map $\psi(x, s) = \psi_0(\phi^m(x), t)$; $y := (\phi^m(x), t)$ and $z := (\phi^m(x), 0)$ are two points of $X \times I$. Hence

$$\begin{aligned} d(h\psi_0(y), h\psi_0(z)) &\leq d(h\psi_0(y), h(\phi^m(x))) + d(h(\phi^m(x)), \psi_0(z)) \\ &\leq 2\|\psi_0\|. \end{aligned}$$

It follows by using Proposition 5.3 that

$$\begin{aligned} d(h\psi(x, s), a) &\leq d(h\phi^m(x), a) + 2\|\psi_0\| \\ &\leq \max(d(h(x), a) - m\varepsilon,\ R + \|\phi\|) + 2\|\psi_0\| \\ &\leq \max(d(h(x), a) - s\varepsilon,\ R + \|\phi\|) + 2\|\psi_0\| + \varepsilon, \end{aligned}$$

so we can choose λ to be $2\|\psi_0\| + \varepsilon$. □

THEOREM 5.6. *If X is contractible with finite cell stabilizers and cocompact n-skeleton then the existence of a finitary contraction $\phi : X^n \to X^n$ towards $a \in M$ implies that X is CC^{n-1} over a with constant lag.*

PROOF. Let $f : S^p \to X^p_{(a,t)}$ be a map, $0 \leq p \leq n - 1$. Since X^{p+1} is p-connected f extends to a map $f_1 : B^{p+1} \to X^{p+1}$. Applying Proposition 5.5, the map $\psi \circ (f \times \mathrm{Id}) : S^p \times [0, \infty) \to X^{p+1}$ can be "glued" to f_1 across $S^p \times \{0\}$ to give a map $f_2 : \mathbb{R}^{p+1} \to X^{p+1}$. Write $\mathbb{R}^{p+1} \times [0, \infty) = \mathbb{R}^{p+2}_+$ and consider $\psi \circ (f_2 \times \mathrm{id}) : \mathbb{R}^{p+2}_+ \to X^{p+2}$. The annulus in $\mathbb{R}^{p+1}$ with inner radius 1 and outer radius r (sufficiently large) adjoined along the sphere of radius r to the hemisphere of radius r in $\mathbb{R}^{p+2}$ gives a $(p+1)$-ball in $\mathbb{R}^{p+2}_+$ whose boundary is $S^p \subseteq \mathbb{R}^{p+1} \subseteq \mathbb{R}^{p+2}_+$. The restriction of $\psi \circ (f_2 \times \mathrm{id})$ to this $(p+1)$-ball $\tilde{B}^{p+1}$ is the required $\tilde{f} : \tilde{B}^{p+1} \to X^{p+1}$ of f. See Figure I. The lag is as λ in Proposition 5.5. □

5.4. Guaranteed shift

By the guaranteed shift towards $a \in M$ of a bounded cellular map $f : D(f) \to X$ on a subcomplex $D(f)$ of X we mean the real number

$$\mathrm{gsh}_a(f) := \inf \mathrm{sh}_{f,a}(D(f)).$$

Observe that $\mathrm{gsh}_a(f)$ is compatible with the G-action in the sense that

$$\mathrm{gsh}_{ga}(gf) = \mathrm{gsh}_a(f)$$

for each $g \in G$. By (5.7) $\mathrm{gsh}_a(f)$ is continuous in the variable a when $D(f)$ is compact.

Let $\mathcal{F} : X \rightsquigarrow X$ be a locally finite homotopically closed G-sheaf. The assumption that $\mathcal{F}$ be homotopically closed guarantees that for every finite subcomplex $K \subseteq X$ there are maps $f \in \mathcal{F}$ with $D(f) = K$ so that we can consider the *maximal guaranteed shift of $\mathcal{F}$ on K* (towards $a \in M$), defined by

$$\mu_a(\mathcal{F}|K) := \max\{\mathrm{gsh}_a(f) | f \in \mathcal{F}, D(f) = K\}.$$

The case when K is the zero skeleton of the carrier of a cell σ of X, $K = C(\sigma)^0$, will be particularly useful to us. We call $\mu_a(\mathcal{F}|C(\sigma)^0)$ the *maximal guaranteed vertex*

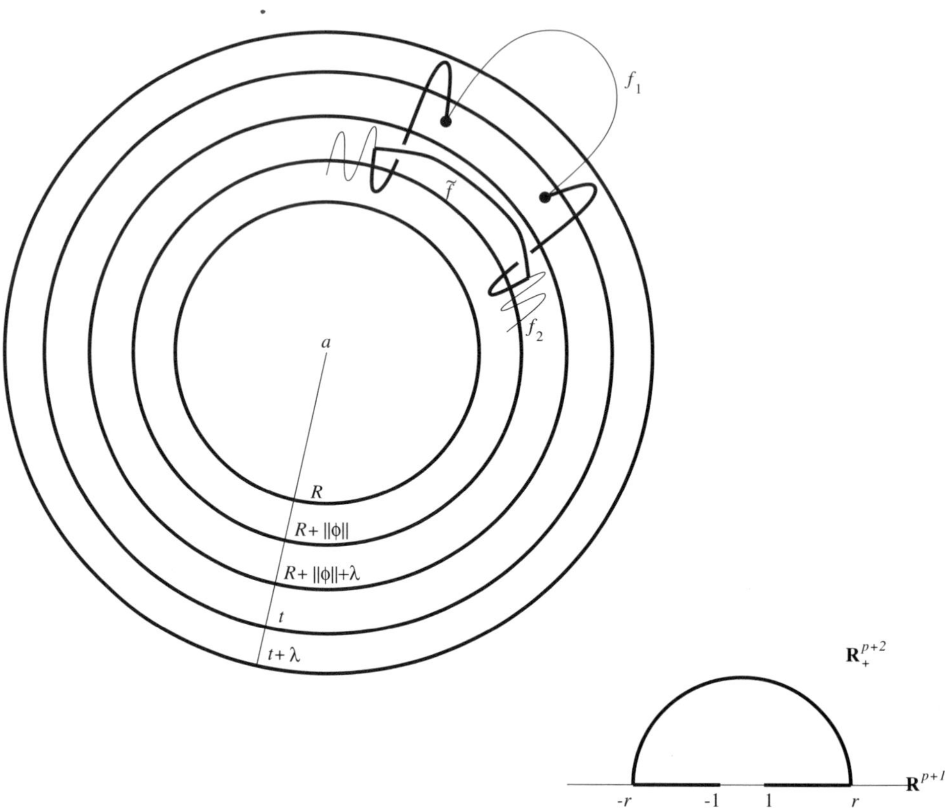

FIGURE 1. Picture for Theorem 5.6

shift on σ. Abusing notation we shorten this to $\mu_a(\mathcal{F}|\sigma)$. As to the G-action we have compatibility in the sense that

$$\mu_{ga}(\mathcal{F}|g\sigma) = \mu_a(\mathcal{F}|\sigma)$$

for all $g \in G, a \in M$. As above, $\mu_a(\mathcal{F} \mid \sigma)$ is continuous in a.

5.5. Defect of a sheaf

Again let $\mathcal{F} : X \rightsquigarrow X$ be a locally finite homotopically closed G-sheaf. Let σ be a cell of X, with $\dim \sigma \geq 1$. We wish to compare the guaranteed shift $\mathrm{gsh}_a(f)$ of a map $f \in \mathcal{F}$ whose domain is the carrier $C(\overset{\bullet}{\sigma})$ with the guaranteed shift of the extensions $\tilde{f} : C(\sigma) \to X$ of f which lie in $\mathcal{F}$: We define the *defect* of $\mathcal{F}$ (towards

a) on σ to be the number

$$d_a(\mathcal{F}|\sigma) = \max_{f\in\mathcal{F}} \min_{\tilde{f}\in\mathcal{F}}\{\mathrm{gsh}_a(f) - \mathrm{gsh}_a(\tilde{f})|D(f)$$
$$= C(\overset{\bullet}{\sigma}), D(\tilde{f}) = C(\sigma), \tilde{f} \text{ extends } f\}.$$

Here again $d_a(\mathcal{F}|\sigma)$ is a shortening of what should be $d_a(\mathcal{F}|C(\sigma))$.

Since $\tilde{f}$ extends f this is a non-negative number. To say that $d_a(\mathcal{F}|\sigma) \leq k$ is to say that for every $f \in \mathcal{F}$ with $D(f) = C(\overset{\bullet}{\sigma})$ there is an extension $\tilde{f}$ of f in $\mathcal{F}$ with $D(\tilde{f}) = C(\sigma)$ such that $\mathrm{gsh}_a(\tilde{f}) \geq \mathrm{gsh}_a(f) - k$. Thus k is an upper bound for the "loss of guaranteed shift" towards a in extending f from $C(\overset{\bullet}{\sigma})$ to $C(\sigma)$ using maps in $\mathcal{F}$; $d_a(\mathcal{F}|\sigma)$ is the best upper bound.

Again, we observe that $d_a(\mathcal{F}|\sigma)$ is compatible with the G-action in the sense that

$$d_{ga}(\mathcal{F}|g\sigma) = d_a(\mathcal{F}|\sigma)$$

for every $g \in G, a \in M$ and all cells σ of X. As above, $d_a(\mathcal{F} \mid \sigma)$ is continuous in a.

Guaranteed vertex shift and defect are used to control the shift towards $a \in M$ of a cross section $\phi : X \to X$ of $\mathcal{F}$, which we construct skeleton by skeleton as follows. For each vertex $v \in X^0$ we choose some $f \in \mathcal{F}$ with $v \in D(f)$ and $\mathrm{sh}_{f,a}(v) = \mathrm{gsh}_a(f|\{v\}) = \mu_a(\mathcal{F}|\{v\})$, and put $\phi_0(v) := f(v)$. This constructs $\phi_0 : X^0 \to X^0$ with the property that $\mathrm{gsh}_a(\phi_0|K^0) = \mu_a(\mathcal{F}|K^0)$ for every finite subset $K^0 \subseteq X^0$. Assume now a cross section ϕ_{k-1} of $\mathcal{F}|X^{k-1}$ has been constructed for some $k \geq 1$. If σ is a k-cell of X we can extend ϕ_{k-1} to the carrier $C(\sigma)$ by a map $\tilde{f} : C(\sigma) \to X$ in $\mathcal{F}$ with $\mathrm{gsh}_a(\tilde{f}) \geq \mathrm{gsh}_a(\phi_{k-1}|C(\overset{\bullet}{\sigma})) - d_a(\mathcal{F}|\sigma)$. This constructs a cross section ϕ_k of $\mathcal{F}|X^k$ with the property that for each finite subcomplex $K \subseteq X^k$

$$\mathrm{gsh}_a(\phi_k|K) \geq \mathrm{gsh}(\phi_{k-1}|K \cap X^{k-1}) - \max\{d_a(\mathcal{F}|\sigma)|\sigma \text{ a } k\text{-cell of } K\},$$

hence,

$$\mathrm{gsh}_a(\phi_k|K) \geq \mu_a(\mathcal{F}|K^0) - \sum_{j=1}^{k} \max\{d_a(\mathcal{F}|\sigma)|\sigma \text{ a } j\text{-cell of } K\}.$$

For every cell σ of X we define the *total defect of $\mathcal{F}$ on σ* (towards $a \in M$) to be

$$\delta_a(\mathcal{F}|\sigma) := \sum_{j=1}^{\dim\sigma} \max\{d_a(\mathcal{F}|\tau)|\tau \text{ a } j\text{-cell of } C(\sigma)\}.$$

If $\dim\sigma = 0$ this means that $\delta_a(\mathcal{F}|\sigma) = 0$. We have proved

PROPOSITION 5.7. *Let $a \in M$. Then every locally finite homotopically closed sheaf $\mathcal{F} : X \rightsquigarrow X$ has a cross section $\phi : X \to X$ with*

$$\mathrm{sh}_{\phi,a}(x) \geq \mu_a(\mathcal{F}|\sigma) - \delta_a(\mathcal{F}|\sigma),$$

for each cell σ of X and each $x \in \sigma$. □

REMARK 5.8. In the construction proving Proposition 5.7 we have not used the G-action. But if $\mathcal{F}$ is, in fact, a G-sheaf (so that ϕ is finitary) and the stabilizer of $a \in M$, denoted by G_a, acts freely on X then in each step of the construction the maps ϕ_k and hence also the cross section ϕ in Proposition 5.7 can be chosen to be G_a-maps.

To make Proposition 5.7 useful we must arrange for $\mu_a(\mathcal{F}|\sigma) - \delta_a(\mathcal{F}|\sigma)$ to be positive.

CHAPTER 6

Construction of Sheaves with Positive Shift

In this section (M, d) is a proper unique-geodesic metric G-space (the need for CAT(0) first appears in §6.3) and X is a G-CW-complex with a control function $h : X \to M$. In view of Proposition 5.7. we wish to construct locally finite closed G-sheaves $\mathcal{F} : X \rightsquigarrow X$ with the property that there are numbers $R \geq 0, \varepsilon > 0$ with

$$\mu_a(\mathcal{F}|\sigma) - \delta_a(\mathcal{F}|\sigma) \geq \varepsilon$$

for every cell σ of X with $h(\sigma) \cap B_R(a) = \emptyset$. Proposition 5.7. will then show that the sheaf $\mathcal{F}$ admits a cross section $\phi : X \to X$ which is a contraction in the sense of §5.3.

6.1. The case when dim $X = 0$

THEOREM 6.1. *Assume $X = X^0$ is a discrete G-set with finite stabilizers and finitely many orbits. Then the following conditions are equivalent.*

(i) *The G-action on M is cocompact*[1].

(ii) *For every $a \in M$ there is a G-finitary contraction $\phi_a : X \to X$ with event radius R independent of a.*

(iii) *There are numbers $R \geq 0$ and $\alpha > 0$ with the property that for each $a \in M$ one can construct a locally finite G-sheaf $\mathcal{F} : X \rightsquigarrow X$ satisfying $\mu_a(\mathcal{F}|\{x\}) \geq \alpha$ for all $x \in X$ with $d(h(x), a) \geq R$.*

(iv) *For any given $\alpha > 0$ there is some $R = R(\alpha)$ such that (iii) holds.*

PROOF. The implication (ii) $\Rightarrow$ (i) is covered by Proposition 5.2. The implication (iii) $\Rightarrow$ (ii) is covered by Proposition 5.7. As (iv) $\Rightarrow$ (iii) is trivial it remains to prove (i) $\Rightarrow$ (iv).

Let $F \subseteq X$ be a fundamental domain and $s \geq 0$ a radius with the property that $GB_s(b) = M$ for every choice of $b \in M$. By replacing F by a G-translate and increasing s if necessary we may assume $a \in h(F) \subseteq B_s(a)$.

Let $R = 3s + \alpha$, and let $S_{R-S}(a)$ denote the sphere of radius $R - s$ around $a \in M$. For each $u \in S_{R-S}(a)$ we choose a point $x_u \in X$ with $d(u, h(x_u)) < s$; this is possible since each point $m \in M$ has distance at most s from some point of $h(X)$. Since $d(u, h(x_u)) < s$, $d(v, h(x_u)) < s$ for all points v in a small neighborhood of u. By compactness of $S_R(a)$ we can thus arrange that all elements x_u are contained in finite subset $W \subseteq X$.

As F and all stabilizers of elements of F are finite, $H := \{g \in G | gF \cap F \neq \emptyset\}$ is a finite subset of G. Now, let $\mathcal{F}_0 : F \rightsquigarrow X$ be the sheaf generated by all maps $f : \{v\} \to X$ with $v \in F$ and $f(v) \in HW$. $\mathcal{F}_0$ is finite and G-saturated (in the sense of §4.2). Hence, by Propositions 4.1 and 4.3, the G-sheaf $\mathcal{F} := G\mathcal{F}_0$ is locally finite. If $x \in X$ and $d(a, h(x)) \geq R$ then we pick $g \in G$ with $gx \in F$ and consider

[1] Recall that by Proposition 3.2 this is equivalent to CC^{-1}

the geodesic segment L from a to ga. One checks that $d(ga,a) \geq R - s$; hence L intersects $S_{R-S}(a)$ in a unique point u. Let $f(gx) := x_u$. This defines a member $f \in \mathcal{F}_0$, and its translate $g^{-1}f$ takes x to $g^{-1}x_u$. We have

$$\begin{aligned} sh_{g^{-1}f,a}(x) &= d(h(x),a) - d(h(g^{-1}x_u),a) \\ &= d(gh(x),ga) - d(h(x_u),ga) \\ &\geq (d(a,ga) - s) - (d(u,ga) + s) \\ &= d(a,ga) - s - (d(a,ga) - d(u,a)) - s \\ &= d(u,a) - 2s \\ &= R - 3s = \alpha. \end{aligned}$$

This shows that $\mu_a(\mathcal{F}|\{x\}) \geq \alpha$. □

6.2. Measuring the loss of guaranteed shift in an extension

In this subsection we assume $n \geq 1$ and that X be CC^{n-1} over some point of M. Then we consider an n-cell σ of X and a cellular map $f : C(\overset{\bullet}{\sigma}) \to X$ and we try to extend f to a map $\tilde{f} : C(\sigma) \to X$ with some control on the shift $\mathrm{gsh}_b(\tilde{f})$ for $b \in M$.

PROPOSITION 6.2. *If X is CC^{n-1} over $b \in M$ then there exists $\lambda_0 \geq 0$ and $\tilde{f} : C(\sigma) \to X$ such that $gsh_b(\tilde{f}) \geq gsh_b(f) - \mathrm{diam}\ hC(\sigma) - \lambda_0$.*

PROOF. Let $u \in C(\overset{\bullet}{\sigma})$ be a point with the property that $d(hf(u),b) =: r_0$ is the maximum distance of any point of $hfC(\overset{\bullet}{\sigma})$ from b. The CC^{n-1} hypothesis gives a lag $\lambda(b,r)$. Let $\lambda_0 := \lambda(b,r_0)$. There exists $\tilde{f} : C(\sigma) \to X$ with $h\tilde{f}(C(\sigma)) \subseteq B_{r+\lambda_0}(b)$. We then have for each $x \in C(\sigma)$

$$\begin{aligned} d(h(x),b) - d(h\tilde{f}(x),b) &\geq d(h(u),b) - \mathrm{diam}\ hC(\sigma) - d(hf(u),b) - \lambda_0 \\ &\geq \mathrm{gsh}_b(f) - \mathrm{diam}\ hC(\sigma) - \lambda_0, \end{aligned}$$

hence the result. □

The same map $\tilde{f}$ works for points b' near b using continuity of gsh_b in the variable b: given any $\varepsilon > 0$ there exist $\delta > 0$ such that

$$\mathrm{gsh}_{b'}(\tilde{f}) > \mathrm{gsh}_{b'}(f) - \mathrm{diam}\ hC(\sigma) - \lambda_0 - \varepsilon$$

for all $b' \in B_\delta(b)$.

We can now let b range over a compact subset B of M. Since B has finite diameter, there is a lag $\lambda = \lambda(r)$ such that X is CC^{n-1} over every point of B with lag λ. A standard compactness argument (pass to a finite subcover of the cover $\{B_\delta(b)\}$ of B) then shows

COROLLARY 6.3. *If $B \subseteq M$ is a compact subset such that X is CC^{n-1} over each $b \in B$ then, given any $\varepsilon > 0$, a finite set S of cellular maps $C(\sigma) \to X$ extending f and a number $\lambda_1 \geq 0$ can be chosen in such a way that for each $b \in B$ there exists $\tilde{f} \in S$ with*

$$gsh_b(\tilde{f}) > gsh_b(f) - diam\ hC(\sigma) - \lambda_1 - \varepsilon.$$

□

6.3. Imposing CAT(0)

We remain in the situation of Section 6.2: We are given a cellular map $f : C(\overset{\bullet}{\sigma}) \to X$ which we try to extend to $\tilde{f} : C(\sigma) \to X$ keeping $\mathrm{gsh}_a(\tilde{f})$ under control. But we need control over $\mathrm{gsh}_a(\tilde{f})$ for all points $a \in M$ and so Corollary 6.3 falls short. In order to improve it we have to impose the assumption that the metric space (M, d) be CAT(0). The CAT(0)-condition first appears[2] via the following lemma ; see the proof of [**BrHa**, III 2.8]:

LEMMA 6.4. *Let (M, d) be a CAT(0) space and let $\varepsilon > 0$. For any $c \in M$ and $r \geq 0$, any number $R > r(1 + 2r/\varepsilon)$ has the property that when $a \in M$ is outside $B_R(c)$ and both p and p' are in $B_r(c)$ then*

$$|(d(a,p) - d(a,p')) - (d(b,p) - d(b,p'))| < \varepsilon,$$

where b is the point on the geodesic segment from c to a distant R from c. □

We apply Lemma 6.4 in the situation of Corollary 6.3 with the following careful choice of $B \subseteq M$. Choose a center $c \in h(\sigma)$ and a radius $r \geq 0$ such that the ball $B_r(c)$ contains both $hC(\sigma)$ and $hf(C(\overset{\bullet}{\sigma}))$. Let R be the radius given by Lemma 6.4 and put $B := B_R(c)$.

To improve on Corollary 6.3 we consider a point $a \in M$ outside B. Let $b \in M$ denote the point on the geodesic segment from c to a with $d(c, b) = R$. Then $b \in B$ so that Corollary 6.3 applies. It guarantees an extension $\tilde{f} \in S$ with the property that for each $x \in C(\sigma)$,

$$\begin{aligned} d(h\tilde{f}(x), b) &< d(h(x), b) - \mathrm{gsh}_b(f) + \mathrm{diam}\ hC(\sigma) + \lambda_1 + \varepsilon \\ &\leq d(c, b) - \mathrm{gsh}_b(f) + 2\mathrm{diam}\ hC(\sigma) + \lambda_1 + \varepsilon. \end{aligned}$$

Hence

$$\begin{aligned} d(h\tilde{f}(x), a) &\leq d(h\tilde{f}(x), b) + d(b, a) \\ &< d(c, a) - \mathrm{gsh}_b(f) + 2\mathrm{diam}\ hC(\sigma) + \lambda_1 + \varepsilon. \end{aligned}$$

On the other hand,

$$d(h(x), a) \geq d(c, a) - \mathrm{diam}\ hC(\sigma),$$

hence

$$d(h(x), a) - d(h\tilde{f}(x), a) > \mathrm{gsh}_b(f) - 3\mathrm{diam}\ hC(\sigma) - \lambda_1 - \varepsilon.$$

As the right hand side of this inequality is independent of $x \in C(\sigma)$ it follows that

$$\mathrm{gsh}_a(\tilde{f}) \geq \mathrm{gsh}_b(f) - 3\mathrm{diam}\ hC(\sigma) - \lambda_1 - \varepsilon.$$

It remains to relate $\mathrm{gsh}_b(f)$ with $\mathrm{gsh}_a(f)$ via Lemma 6.4. For every $y \in C(\overset{\bullet}{\sigma})$ it yields, by putting $p = h(y)$ and $p' = hf(y)$,

$$|\mathrm{sh}_{f,a}(y) - \mathrm{sh}_{f,b}(y)| < \varepsilon.$$

It follows that $\mathrm{gsh}_b(f) \geq \mathrm{gsh}_a(f) - \varepsilon$, hence

$$\mathrm{gsh}_a(\tilde{f}) \geq \mathrm{gsh}_a(f) - 3\mathrm{diam}\ hC(\sigma) - \lambda_1 - 2\varepsilon.$$

Thus we have proved

[2]In Remark 7.8 we indicate the other places in this paper where we use the CAT(0) condition.

PROPOSITION 6.5. *Assume that X is CC^{n-1} over a* CAT(0) *space M. Then, given any $\varepsilon > 0$, a finite set S of cellular maps $C(\sigma) \to X$ extending f and a number $\lambda_1 \geq 0$ can be chosen in such a way that for each $a \in M$ there exists $\tilde{f} \in S$ with*

$$gsh_a(\tilde{f}) \geq gsh_a(f) - 3\text{diam } hC(\sigma) - \lambda_1 - 2\varepsilon.$$

□

6.4. The main technical theorem

We are now in a position to prove the following crucial consequence of the CC^{n-1}-condition over a CAT(0)-space M.

PROPOSITION 6.6. *Let X be a cocompact G-CW-complex with finite stabilizers, let $h : X \to M$ be a control function and let $n \geq 1$. Assume (M, d) is* CAT(0) *and X is CC^{n-1} over M. Then there is a constant $\eta \geq 0$ with the property that every homotopically closed locally finite G-sheaf $\mathcal{F} : X^{n-1} \rightsquigarrow X^{n-1}$ can be embedded in a homotopically closed locally finite G-sheaf $\tilde{\mathcal{F}} : X^n \rightsquigarrow X^n$ with $d_a(\tilde{\mathcal{F}}|\sigma) \leq \eta$ for every $a \in M$ and all n-cells σ of X.*

PROOF. Fix an n-cell σ of X. As $\mathcal{F}$ is locally finite there are only finitely many members $f \in \mathcal{F}$ with $D(f) = C(\overset{\bullet}{\sigma})$. Hence Proposition 6.5 yields a finite set $S(\sigma)$ of cellular maps $C(\sigma) \to X^n$ such that for each $f \in \mathcal{F}$ with $D(f) = C(\overset{\bullet}{\sigma})$ there is some $\tilde{f} \in S(\sigma)$ extending f with $\text{gsh}_a(f) - \text{gsh}_a(\tilde{f}) \leq \Lambda$, where Λ can be chosen independent of σ. $S(\sigma)$ depends on σ; but for the G-translates $g\sigma$ of σ, we do not have to choose $S(g\sigma)$ anew but can put $S(g\sigma) := gS(\sigma)$. This is because $\text{gsh}_{ga}(gf) = \text{gsh}_a(f)$; see §5.4. Let $\tilde{\mathcal{F}}$ denote the G-sheaf generated by $\mathcal{F}$ together with the maps $\tilde{f} \in S(\tau)$, as τ runs through a (finite) set of n-cells representing all G-orbits. By Proposition 4.4 $\tilde{\mathcal{F}}$ fulfills the conclusion of the Proposition. □

COROLLARY 6.7. *Let $h : X \to M$ as in Proposition 6.6. If M is CAT(0) and X is CC^{n-1} over M then there is a homotopically closed locally finite G-sheaf $\mathcal{F} : X^n \rightsquigarrow X^n$, a finite radius $R \geq 0$, and some $\varepsilon > 0$ such that*

$$\mu_a(\mathcal{F}|\sigma) - \delta_a(\mathcal{F}|\sigma) \geq \varepsilon$$

for all $a \in M$ and all cells σ of X^n with $h(\sigma) \cap B_R(a) = \emptyset$.

PROOF. Uniformly CC^{-1} means that the G-action on M is cocompact. Hence Theorem 6.1 applies and yields a radius $R = R(\alpha)$ and a locally finite G-sheaf $\mathcal{F}^0 : X^0 \rightsquigarrow X^0$ satisfying $\mu_a(\mathcal{F}^0|v) \geq \alpha$ for each vertex $v \in X^0$ with $d(h(v), a) \geq R$, where α is an arbitrary positive number yet to be chosen. Now we apply Proposition 6.6 in each dimension to see that $\mathcal{F}^0$ can be embedded in a homotopically closed locally finite G-sheaf $\mathcal{F}$ with a bound on $\delta_a(\mathcal{F}|\sigma)$ independent of $\mathcal{F}^0, \sigma$, and $a \in M$. Choosing α greater than or equal to this bound yields the corollary. □

THEOREM 6.8. *Let $h : X \to M$ be a contractible G-CW-complex over M with finite stabilizers and cocompact n-skeleton. Under the assumption that the control space M is* CAT(0) *the following conditions are equivalent*

(i) *X is CC^{n-1} over M.*
(ii) *For every $a \in M$ there is a G-finitary contraction $\phi_a : X^n \to X^n$ with event radius independent of a.*

(iii) *G acts cocompactly on M and there is a G-finitary contraction $\phi : X^n \to X^n$ towards some $a \in M$.*

(iv) *There is a locally finite homotopically closed G-sheaf $\mathcal{F} : X^n \rightsquigarrow X^n$, a radius $R \geq 0$, and a positive number ε such that*

$$\mu_a(\mathcal{F}|\sigma) - \delta_a(\mathcal{F}|\sigma) \geq \varepsilon$$

for all $a \in M$ and all cells σ of X^n with $h(\sigma) \cap B_R(a) = \emptyset$.

(v) *X is CC^{n-1} over M with constant lag.*

PROOF. As "CC^{-1}" implies "G acts cocompactly on M" the implication (i) $\Rightarrow$ (iv) is covered by Corollary 6.7. The implication (iv) $\Rightarrow$ (ii) is covered by Proposition 5.7. (ii) $\Rightarrow$ (iii) is easy since the assumption (ii) for $n = 0$ implies, by Proposition 5.2, that X is CC^{-1} and hence M is cocompact. Theorem 5.6, when stated for a cocompact G-space M, covers the remaining implication (iii) $\Rightarrow$ (v). And (v) $\Rightarrow$ (i) is trivial. □

CHAPTER 7

Controlled Connectivity as an Open Condition

7.1. The topology on the set of all G-actions

Let (M, d) be a proper metric space and let G be group of type F_n. We consider left actions of G on M where, in contrast to the previous sections, the action is allowed to vary. We give G the discrete topology, and we give $\mathrm{Isom}(M)$ and $\mathrm{Hom}(G, \mathrm{Isom}(M))$ the compact-open topology.

We recall some general topology; for details, see [**Du**], especially Section XII 1.3 and 5.2 and IX 9.2. When A, B and C are topological spaces and C^B denotes the space of all continuous functions $B \to C$ with the compact-open topology, a function $\alpha : A \times B \to C$ is continuous if and only if its adjoint $\hat{\alpha} : A \to C^B$ is continuous, provided B is locally compact. (Note that our M is locally compact.) The space C^B is metrizable provided B and C are 2nd countable, B is locally compact and Hausdorff, and C is regular. (Note that our M is 2nd countable, being metrizable and sigma compact.) In the compact-open topology sequential convergence agrees with uniform convergence on compact sets ([**Du**, XII 7.2]), provided the target space is metrizable. Summarizing:

PROPOSITION 7.1. *$\mathrm{Isom}(M)$ and $\mathrm{Hom}(G, \mathrm{Isom}(M))$ are metrizable 2nd countable spaces. In both function spaces the compact-open topology is the topology of uniform convergence on compact sets. For any space A, a function $A \to M^M$ is continuous if and only if its adjoint function $A \times M \to M$ is continuous.* □

7.2. Continuous choice of control functions

In Section 2.2 we chose a contractible free G-CW complex X with G-finite n-skeleton and a G equivariant control function $h : X \to M$ into the given control space M. Now we vary the G-action $\rho : G \to \mathrm{Isom}(M)$ and choose a (G, ρ)-equivariant control function $h_\rho : X \to M$ for each ρ. In order to emphasize the dependence on ρ we write $\rho(g)a$ for the effect of the action of $g \in G$ on $a \in M$. The assignment $\rho \mapsto h_\rho$ thus defines a function $\mathrm{Hom}(G, \mathrm{Isom}(M)) \to M^{X^n}$.

PROPOSITION 7.2. *If (M, d) is a proper unique-geodesic metric space this function can be chosen to be continuous.*

PROOF. Pick a base point $b \in M$. Pick a representative cell σ for each G-orbit of cells of X, and write $x_\sigma \in X$ for the "barycenter" of σ (i.e., the image of the origin under a characteristic map $\mathbb{B}^k \twoheadrightarrow \sigma$). We will construct $h_\rho : X \to M$ in such a way that $h_\rho(x_\sigma) = b$ for each representative cell σ. By G-equivariance this defines h_ρ on the zero skeleton X^0. Assuming, then, that $h_\rho : X^k \to M$ is already defined on the k-skeleton we extend it to a representative $(k+1)$-cell σ by regarding σ as the mapping cone of its attaching map $S^k \to X^k$. Extend h_ρ to σ by mapping each cone line linearly to the unique geodesic joining the h_ρ-images of its endpoints. Uniqueness of geodesic implies that geodesics vary continuously with their endpoints [**BrHa**, I.3.11], so this extension is continuous. Extend h_ρ to be (G,ρ)-equivariant on X^{k+1}. The map $h_\rho : X \to M$ is thus "canonically defined" and is therefore continuous in ρ. □

For the rest of this section we assume that h_ρ depends continuously on ρ. We now write $\mathrm{sh}^\rho_{f,a} : D(f) \to \mathbb{R}$ in order to emphasize that it depends on the G-action.

PROPOSITION 7.3. *Let $\mathcal{F} : X \rightsquigarrow X$ be a locally finite homotopically closed G-sheaf on a G-CW complex X over M. Then the maximal guaranteed vertex shifts $\mu^\rho_a(\mathcal{F}|\sigma)$ and the defect $d^\rho_a(\mathcal{F}|\sigma)$ are jointly continuous in the variables (ρ, a). Moreover we have*

$$|\mu^\rho_a(\mathcal{F}|\sigma) - \mu^\rho_b(\mathcal{F}|\sigma)\,| \leq 2d(a,b)$$

and

$$|d^\rho_a(\mathcal{F}|\sigma) - d^\rho_b(\mathcal{F}|\sigma)\,| \leq 4d(a,b)$$

for all $a, b \in M$. □

PROOF. This is elementary. It is helpful to use the even more precise notation $\mathrm{sh}^h_{f,a}(x)$, $\mathrm{gsh}^h_a(f)$ etc. where $h : X \to M$ is any map, and to prove joint continuity in h and a using Formula (5.7). □

7.3. Imposing CAT(0)

Condition (iv) of Theorem 6.8 expresses the CC^{n-1}-property of an action $\rho : G \to \mathrm{Isom}(M)$ in terms of the existence of a locally finite homotopically closed G-sheaf $\mathcal{F} : X^n \rightsquigarrow X^n$ whose functions have to satisfy a certain inequality. In order to prove the "openness" of the CC^{n-1}-property we will show that if we fix the G-sheaf $\mathcal{F}$ then this inequality remains true under small perturbation of the action ρ. Proposition 7.3 is the first step in this but is not sufficient because the range of the parameter $a \in M$ in the inequality is not compact. The missing ingredient is a result expressing that the functions of a, $\mu^\rho_a(\mathcal{F}|\sigma)$ and $d^\rho_a(\mathcal{F}|\sigma)$, are, to some extent, ruled by their values on points $a \in M$ within a bounded distance from $h_\rho(\sigma)$. This requires, once again, the assumption that M be CAT(0).

Before we can state the missing result in Proposition 7.5 we need some preparation. We consider a closed locally finite G-sheaf $\mathcal{F} : X \rightsquigarrow X$ on a G-CW complex X over the CAT(0) space M. For each cell σ of X we consider the carrier $C(\sigma)$ and the restricted sheaf $\mathcal{F}|C(\sigma)$. With respect to a given G-action $\rho : G \to \mathrm{Isom}(M), \mathcal{F}|C(\sigma)$ has a well defined norm over M in the sense of §5.1. We put $r = r_\sigma(\rho) := \|\mathcal{F}|C(\sigma)\| + \mathrm{diam}\; h_\rho C(\sigma)$. Let

$$L_\sigma : \mathrm{Hom}(G, \mathrm{Isom}(M)) \times \mathbb{R}_{>0} \to \mathbb{R}, \tag{7.1}$$

be defined by $L_\sigma(\rho, \varepsilon) = r + 2r^2/\varepsilon$. Observe that joint continuity (resp. G-equivariance) of $h_\rho(x)$ in ρ and x implies that the norm of $\mathcal{F}|C(\sigma)$ (which depends on ρ) and the diameter of $h_\rho C(\sigma)$ are continuous in ρ (resp. G-invariant), hence

PROPOSITION 7.4. *L_σ is continuous, and $L_{g\sigma} = L_\sigma$ for every $g \in G$ and every cell σ of X.*

□

In order to keep control over distances between points of M and h_ρ-images of cells we choose a "barycenter" $x_\sigma \in \sigma$ for every cell σ of X in such a way that $x_{g\sigma} = gx_\sigma$ for each $g \in G$. Let $c_\sigma := h_\rho(x_\sigma)$. Then $c_{g\sigma} = \rho(g)c_\sigma$ for each $g \in G$.

PROPOSITION 7.5. *Assume M is* CAT(0). *Fix a cell σ of X, a number $\varepsilon > 0$ and a G-action $\rho : G \to \mathrm{Isom}(M)$. Then given any radius $R > L_\sigma(\rho, \varepsilon)$ and a point $a \in M$ with $d(c_\sigma, a) \geq R$ we have*

$$|\mu_a^\rho(\mathcal{F}|\sigma) - \mu_b^\rho(\mathcal{F}|\sigma)| < \varepsilon, \tag{7.2}$$

in general, and

$$|d_a^\rho(\mathcal{F}|\sigma) - d_b^\rho(\mathcal{F}|\sigma)| < 2\varepsilon \tag{7.3}$$

when $\dim \sigma > 0$, *where $b = b_\sigma(\rho, a)$ stands for the point on the geodesic segment from c_σ to a at distance R from c_σ.*

PROOF. The function $r = r_\sigma(\rho)$ used in (7.1) was chosen so that the ball $B_r(c_\sigma)$ contains both $h_\rho C(\sigma)$ and $h_\rho f(C(\sigma))$ for each $f \in \mathcal{F}|C(\sigma)$. In this situation we can apply Lemma 6.4 with $p = h_\rho(x), p' = h_\rho f(x)$ and find, for each $x \in D(f)$,

$$|\mathrm{sh}_{f,a}^\rho(x) - \mathrm{sh}_{f,b}^\rho(x)\,| < \varepsilon. \tag{7.4}$$

(7.2) is an immediate consequence of (7.4). As to (7.3) we first observe that (7.4) implies the corresponding inequality,

$$|\mathrm{gsh}_a(f) - \mathrm{gsh}_b(f)\,| < \varepsilon,$$

for the guaranteed shifts of f. Plugging this twice into the definition of the defect $d_a(\mathcal{F} \mid \sigma)$ in §5.5 yields the required inequality (7.3). □

For later use we also record

PROPOSITION 7.6. *In the situation of Proposition 7.5 the points $b = b_\sigma(\rho, a)$ satisfy the inequality*

$$d(b_\sigma(\rho, a), b_\sigma(\rho', a)) \leq d(c_\sigma(\rho), c_\sigma(\rho'))$$

for any two actions $\rho, \rho' : G \to$ Isom(M).

PROOF. By the definition of CAT(0) it suffices to prove the inequality in the Euclidean plane. Here it is elementary – e.g. use the Cosine Theorem to compute the two distances. □

7.4. The Openness Theorem

We are now in a position to prove our main result.

THEOREM 7.7. *For isometric actions ρ of a group G of type F_n on a proper* CAT(0) *space M, controlled n-connectedness is an open condition; i.e., if ρ is CC^{n-1} over M there is a neighborhood N of ρ in* $\mathrm{Hom}(G, \mathrm{Isom}(M))$ *such that every $\rho' \in N$ is also CC^{n-1} over M.*

PROOF. Let X be the universal cover of a $K(G,1)$-complex with finite n-skeleton. Make a continuous choice of control functions $h_\rho : X^n \to M$, fixing a base point $*$, and of barycenters x_σ as in the proof of Proposition 7.2. Let T be a set of representatives of the G-orbits of the cells of X. We have $h_\rho(x_\sigma) = *$ for all $\sigma \in T$. There is a finite set $H \subseteq G$ such that HT contains all cells of the carriers $C(\sigma)$ with $\sigma \in T$.

Now let ρ be a G-action on M which is CC^{n-1}. Let $\mathcal{F} : X^n \rightsquigarrow X^n, R \geq 0$ and $\varepsilon > 0$ be as given in Condition (iv) of Theorem 6.8. We fix a natural number m which we will specify later. As we can replace R by a larger number, if necessary, we may assume that $R > L_\sigma(\rho, \frac{\varepsilon}{m})$.

Now we consider the following subsets $N_i(\sigma) \subseteq \mathrm{Hom}(G, \mathrm{Isom}(M))$.

$$\begin{aligned} N_1(\sigma) &:= \{\rho' \big| L_\sigma(\rho', \frac{\varepsilon}{m}) < R\} \\ N_2(\sigma) &:= \{\rho' \big| |\mu_b^\rho(\mathcal{F}|\sigma) - \mu_b^{\rho'}(\mathcal{F}|\sigma)\big| < \frac{\varepsilon}{m}, \, b \in B_R(*)\} \\ N_3(\sigma) &:= \{\rho' \big| |d_b^\rho(\mathcal{F}|\sigma) - d_b^{\rho'}(\mathcal{F}|\sigma)\big| < \frac{\varepsilon}{m}, \, b \in \rho(H)B_R(*)\} \\ \text{and } N_4 &:= \{\rho' \big| d(\rho(g)*, \rho'(g)*) < \frac{\varepsilon}{m}, \, g \in H\} \end{aligned}$$

These sets are open by Propositions 7.3 and 7.4 since H is finite and $B_R(*)$ is compact. Since T and HT are finite sets of cells the intersection

$$N := \bigcap_{\sigma \in T} (N_1(\sigma) \cap N_2(\sigma)) \cap (\bigcap_{\tau \in HT} N_3(\tau)) \cap N_4$$

is also open. Of course, $\rho \in N$.

Let $a \in M$ with $d(\rho(g)*, a) \geq R + \frac{\varepsilon}{m}$ for all $g \in H$. This implies that for each $\tau = g\sigma \in HT$ not only the element $b_\tau(\rho, a)$ occurring in Proposition 7.5 but also $b_\tau(\rho', a)$ for all $\rho' \in N_4$ is well defined. Indeed, this follows since $c_\tau(\rho) = c_{g\sigma}(\rho) = \rho(g)c_\sigma(\rho) = \rho(g)*$, and similarly for ρ'.

For all $\rho' \in N$ and all $\sigma \in T$ we find

$$\begin{aligned} |\mu_a^\rho - \mu_a^{\rho'}| &\leq |\mu_a^\rho - \mu_b^\rho| + |\mu_b^\rho - \mu_b^{\rho'}| + |\mu_b^{\rho'} - \mu_a^{\rho'}| \\ &< \frac{\varepsilon}{m} + \frac{\varepsilon}{m} + \frac{\varepsilon}{m} = \frac{3\varepsilon}{m}. \end{aligned}$$

Here we have omitted the argument $\mathcal{F}|\sigma$ and we have used b to denote the point $b = b_\sigma(\rho, a)$; we have used $c_\sigma(\rho) = * = c_\sigma(\rho')$ so that $b_\sigma(\rho, a) = b_\sigma(\rho', a)$ and have applied Proposition 7.5 both to ρ and ρ'; and we have used $b \in B_R(*)$ so that the definition of $N_2(\sigma)$ applies.

The corresponding computation for d_a^ρ instead of μ_a^ρ is more subtle since we will need it not only for $\sigma \in T$ but for all $\tau = g\sigma \in HT$ in order to get control over the total defect δ_a^ρ. Again omitting the argument $\mathcal{F}|g\sigma$ we find the inequality

$$|d_a^\rho - d_a^{\rho'}| \leq |d_a^\rho - d_b^\rho| + |d_b^\rho - d_{b'}^{\rho'}| + |d_{b'}^{\rho'} - d_a^{\rho'}|,$$

where $b = b_{g\sigma}(\rho, a)$ and $b' = b_{g,\sigma}(\rho', a)$. By Proposition 7.5 the first and third terms of the right hand side are smaller than $\frac{2\varepsilon}{m}$. But the middle term needs more care. We write

$$|d_b^\rho - d_{b'}^{\rho'}| \leq |d_b^\rho - d_b^{\rho'}| + |d_b^{\rho'} - d_{b'}^{\rho'}|$$

and observe that $b = b_{g\sigma}(\rho, a) \in B_R(c_{g\sigma}(\rho)) = \rho(g)B_R(*)$. Hence the definition of $N_3(g\sigma)$ applies and shows that the first terms of the right hand side is less than $\frac{\varepsilon}{m}$. The second term is at most equal to $4d(b, b')$ by Proposition 7.3 which is less than or equal to $4d(c_{g\sigma}(\rho), c_{g\sigma}(\rho')) = 4d(\rho(g)*, \rho'(g)*) \leq \frac{4\varepsilon}{m}$, by Proposition 7.6.

Thus we have proved that for the defect on a cell $\tau \in HT$ we have

$$\left|d_a^\rho(\mathcal{F}|\tau) - d_a^{\rho'}(\mathcal{F}|\tau)\right| < \frac{9\varepsilon}{m}.$$

This is enough information to obtain a similar formula for the total defect on the cells $\sigma \in T$. Indeed, we find

$$\left|\delta_a^\rho(\mathcal{F}|\sigma) - \delta_a^{\rho'}(\mathcal{F}|\sigma)\right| < \frac{9\varepsilon \cdot \dim \sigma}{m} \leq \frac{9\varepsilon n}{m}$$

Putting things together we find that if $\rho' \in N$ then we have on each cell $\sigma \in T$

$$\left|(\mu_a^\rho - \delta_a^\rho) - (\mu_a^{\rho'} - \delta_a^{\rho'})\right| < \frac{3(3n+1)\varepsilon}{m}.$$

Now choose $m := 6(3n+1)$ so that, since ρ satisfies (iv) of Theorem 6.8,

$$\left|\mu_a^{\rho'}(\mathcal{F}|\sigma) - \delta_a^{\rho'}(\mathcal{F}|\sigma)\right| > \frac{\varepsilon}{2}$$

for all $\sigma \in T$. Our computation applies for all $a \in M$ with $d(c_\sigma(\rho'), a) = d(*, a) \geq R_0$, where $R_0 = R + \frac{\varepsilon}{m} + \max d(\rho(F)*, *)$. All cells are translates of cells in T, so that the G-compatibility conditions in §§5.4 and 5.5 imply that condition (iv) of Theorem 6.8 holds true for ρ'. □

REMARK 7.8. The case $n = 0$ – i.e. the condition that M be cocompact – is of course simpler. First, it requires Theorem 6.1 rather than Theorem 6.8, and Theorem 6.1 holds true without the assumption that M be CAT(0). Secondly, the defect function d_a^ρ plays no role in dimension 0, so the more subtle technicalities of the above proof disappear – in particular Proposition 7.6, which was another instance where the CAT(0) condition appeared, is not needed. However, the case $n = 0$ does require the first part of Proposition 7.5 which was the third instance where CAT(0) was needed. We do not know to what extent an openness theorem for cocompact actions on geodesic metric spaces beyond the CAT(0) case holds.

The case $n = 0$ combined with Proposition 3.2 gives:

COROLLARY 7.9. *For isometric actions of a group G on a proper* CAT(0) *space M, cocompactness is an open condition.* □

REMARK 7.10. If M is a simply connected manifold of non-positive sectional curvature, then $\text{Isom}(M)$ is a Lie group. By a theorem of Weil [**We 60**], it follows that if $\rho : G \to \text{Isom}(M)$ is injective and acts properly discontinuously and cocompactly on M then there is a compact fundamental domain $K \subset M$ for ρ and a neighborhood N of ρ in $\text{Hom}(G, \text{Isom}(M))$ such that every $\rho' \in N$ is injective and acts properly discontinuously, and K is also a fundamental domain for ρ'. This is a stronger conclusion than that of Corollary 7.9, but under much more restrictive hypotheses.

CHAPTER 8

Completion of the proofs of Theorems A and A′

8.1. Controlled acyclicity

In the set-up of §2.3, we say X is *controlled* $(n-1)$*-acyclic* (CA^{n-1}) over a (with respect to h) if for all $r \geq 0$ and $-1 \leq p \leq n-1$ there exists $\lambda \geq 0$ such that every $\mathbb{Z}$-cycle in $X_{(a,r)}$ bounds in $X_{(a,r+\lambda)}$.

8.2. The F_n-Criterion

In [**Br 87$_\mathbf{I}$**, Theorem 2.2] Brown gives an FP_n-Criterion analogous to the F_n-Criterion: i.e. F_n is replaced by FP_n and CC^{n-1} is replaced[1] by CA^{n-1}. From this, Brown deduces the "if" direction of the F_n-Criterion by proving it for $n \leq 2$: this together with the FP_n-Criterion gives "F_2 and FP_n" which is well known to be equivalent to F_n. He also proves the "only if" direction of the F_2-Criterion. To complete the proof of the "only if" direction of the F_n-Criterion via the FP_n-Criterion one needs a "Hurewicz-type" theorem for which we do not know a reference, but which will appear in [**Ge**]:

THEOREM 8.1. *Let $(Y_r)_{r\in\mathbb{R}}$ be a filtration of the CW complex Y by subcomplexes, and let $f : Y \to \mathbb{R}$ be the control function $f(y) = \inf\{r \mid y \in Y_r\}$. With respect to f, Y is CC^{n-1} if and only if Y is CC^1 and CA^{n-1}.* □

8.3. Proof of Theorem A

It only remains to prove:

LEMMA 8.2. *If the G-action on M has discrete orbits then the G_a-complexes $X^n_{(a,r)}$ are cocompact.*

PROOF. The cells of $X^n_{(a,r)}$ decompose into finitely many classes under the G-action, and those are, of course, G_a-invariant. We claim that for every cell σ of $X^n_{(a,r)}$ the set of all cells of the form $g\sigma \in X^n_{(a,r)}$, $g \in G$, consists of only finitely many G_a-orbits.

For every cell σ of X^n we choose a "barycenter" $x_\sigma \in \sigma$ in such a way that $x_{g\sigma} = gx_\sigma$ for all $g \in G$. Since G acts discretely on M and X^n has only finitely many G-orbits of cells, $B := \{h(x_\sigma) | \sigma \subset X^n\}$ is a discrete subset of M. Since $B_r(a)$ is compact the subset $C := B \cap B_r(a)$ is finite. For each $c \in C$ the stabilizers G_a and G_c are commensurable and we choose a (finite) set of coset representatives for $(G_a \cap G_c)\backslash G_c$. Putting these together, we get a finite subset $L \subset G$ with $G_c \subset G_a L$, for all $c \in C$. Moreover, we consider a second finite subset $T \subset G$ by choosing, for each pair of elements $c_1, c_2 \in C$ which are in the same G-orbit an element $t \in G$ with $g_2 = tg_1$.

[1]In Brown's terminology the corresponding homotopy or homology is "essentially trivial" in dimensions $\leq n-1$.

We complete the proof of the Claim by showing that if $g \in G$ and σ is a cell of X^n such that both σ and $g\sigma$ are in $X_{(a,r)}$ then $g \in G_a LT$. Indeed, both $c := h(x_\sigma)$ and $gc = h(x_{g\sigma})$ are in C; hence there is some $t \in T$ with $gc = tc$. This shows that $gt^{-1} \in G_{tc}$, hence $g \in G_{tc}T$. But $tc \in C$, so that $G_{tc} \subseteq G_a L$. This proves the Claim. □

8.4. Properly discontinuous actions

Let (M, d) be a metric space and Q a group which acts on M by isometries. Recall that this action is properly discontinuous if every point of M has a neighborhood U such that $\{q \in Q | U \cap qU \neq \emptyset\}$ is finite. To complete Footnote 8 (and hence the proof that Theorem A implies Theorem A′) we only need:

LEMMA 8.3. *If M is locally compact then the Q-action on M is properly discontinuous if and only if its orbits are discrete and its point stabilizers are finite.*

PROOF. For each $a \in M$ and $\varepsilon > 0$ we consider the subset $L(a, \varepsilon) \subset Q$, $L(a, \varepsilon) := \{q | d(a, qa) < \varepsilon\}$. Observe that $L(a, \varepsilon)$ contains the stabilizer Q_a. On the one hand, to say that the orbit Q_a is a closed discrete subspace of M is to say that there is some $\varepsilon > 0$ with $L(a, \varepsilon) = Q_a$. On the other hand, the definition of "properly discontinuous", when translated into a statement about the metric d , says that for every $a \in M$ there is some $\varepsilon > 0$ such that $L(a, \varepsilon)$ is finite. But then one can take $0 < \nu < \varepsilon$ with $d(a, qa) > \nu$, for all $q \in L(a, \varepsilon)$ with $qa \neq a$, and one has $Q_a = L(a, \nu)$ finite for some $\nu > 0$. This proves the Lemma. □

CHAPTER 9

The Invariance Theorem

Here we prove the Invariance Theorem 3.3 which says that in checking the CC^{n-1} property over $a \in M$ we may use any $a \in M$, any n-dimensional, $(n-1)$-connected, cocompact rigid G-CW complex Y and any G-map $h : Y \to M$, provided only that the stabilizer of each p-cell of Y is of type $F_{n-p}, 0 \leq p \leq n-1$. Recall that in Chapter 3 we have already proved Invariance among such complexes Y which are free: we will use that in the present proof.

Independence of a is clear. Given Y, independence of h is proved as in Proposition 3.1. To establish independence of Y, we let X be a contractible free G-CW complex. Then $X \times Y$ is an $(n-1)$-connected free G-CW complex. We form a commutative diagram:

$$\begin{array}{ccccccc} \tilde{W} & \xrightarrow{\tilde{\alpha}} & X \times Y & \xrightarrow{\tilde{\beta}} & Y & \xrightarrow{h} & M \\ \downarrow & & \downarrow & & \downarrow & & \\ W & \xrightarrow{\alpha} & G\backslash(X \times Y) & \xrightarrow{\beta} & G\backslash Y & & \end{array}$$

The space $G\backslash(X \times Y)^n$ is the n-skeleton of a $K(G,1)$-complex. The G-map $\tilde{\beta}$ is projection on the Y-factor. There is a cellular map β making the square involving β and $\tilde{\beta}$ (in the diagram) commute.

Consider a cell σ of $G\backslash Y$. By rigidity it is the homeomorphic image of a cell of Y, and if $z \in \overset{\circ}{\sigma}$ then $Z(\sigma) := \beta^{-1}(z)$ is the n-skeleton of a $K(G_\sigma, 1)$-complex (though not a subcomplex of $G\backslash(X \times Y)$ when dim $\sigma > 0$). Indeed, $G\backslash(X \times Y)$ can be decomposed[1] as a "CW complex of CW complexes", parametrized by the cells σ of $G\backslash Y$, in which each complex $C(\sigma) := Z(\sigma) \times B^{\dim(\sigma)}$ is glued to $\beta^{-1}(G\backslash Y^{\dim(\sigma)-1})$ along $Z(\sigma) \times S^{\dim(\sigma)-1}$. It follows that, for each $y \in Y$, $\tilde{\beta}^{-1}(y)$ an n-dimensional $(n-1)$-connected G_y-complex.

Replacing the complex $Z(\sigma)$ by a new complex $Z'(\sigma)$ of the same homotopy type but with finite $(n - \dim(\sigma))$-skeleton, one gets W, again the n-skeleton of a $K(G,1)$-complex, by gluing the complexes $C'(\sigma) := Z'(\sigma) \times B^{\dim(\sigma)}$, as before, by induction on dim σ. The map α naturally arising in this construction takes each $C'(\sigma)$ to $C(\sigma)$. Since $G\backslash Y$ is a finite complex, one sees by counting cells that W is a finite complex. The commutative diagram is completed with the universal cover $\tilde{W}$ of W and the lift $\tilde{\alpha}$ of α. The map $\tilde{\alpha}$ is a G-map, and for each $y \in Y$ $\tilde{\alpha}^{-1}\tilde{\beta}^{-1}(y)$ is n-dimensional and $(n-1)$-connected. It follows by the Vietoris-Smale Theorem [**Sm 57**] that for every subcomplex K of Y, $\tilde{\beta} \circ \tilde{\alpha} \mid : \tilde{\alpha}^{-1}\tilde{\beta}^{-1}(K) \to K$ induces an

[1]Compare the "graphs of complexes" in [**SW 79**, p. 165 et seq.] which the present argument generalizes, and see [**Ge**, Ch 2] or [**Sta 98**, §3] for details omitted here. This is in the spirit of "complexes of groups" as expounded in [**BrHa**, Ch III $\mathcal{C}$].

isomorphism on homotopy groups in dimensions $\leq n-1$. Thus Y is CC^{n-1} over a with respect to the control function h if and only $\tilde{W}$ is CC^{n-1} over a with respect to $h \circ \tilde{\beta} \circ \tilde{\alpha}$. But $\tilde{W}$ is a free G-complex. Since invariance has already been proved for free G-complexes, we are done.

Part 2

The geometric invariants

Short summary of Part 2

Let G be a group, let M be a simply connected "non-positively curved", i.e. CAT(0), metric space, and let $\rho : G \to \text{Isom}(G)$ be an action of G on M by isometries. In Part I we introduced, when G is of type F_n, a property of the action ρ which we called "controlled $(n-1)$-connectedness", abbreviated CC^{n-1}. This property was defined in terms of the filtration of M by the balls $B_r(a)$ centered at some base point $a \in M$ together with a free contractible G-CW-complex over M. In the present paper we pay attention to the points at infinity, i.e. the "boundary" ∂M of the space M. For each point $e \in \partial M$ we introduce a new property of the action ρ, analogous to CC^{n-1} but defined using the filtration of M by horoballs centered at e rather than by the balls $B_r(a)$; we call this "CC^{n-1} over e". Whereas in the previous situation the property CC^{n-1} was independent of the base point $a \in M$ it will now depend in a delicate way upon the point $e \in \partial M$. Therefore the subset of ∂M

$$\Sigma^n(\rho) := \{e \in \partial M \mid \rho \text{ is } CC^{n-1} \text{ over } e\},$$

becomes interesting in its own right, and we investigate its structure.

The isometric action ρ of G on M induces a topological action of G on ∂M when ∂M is given the (compact) "cone" topology[2], and $\Sigma^n(\rho)$ is invariant with respect to this action. However, it may happen that the closure, $\text{cl}_{\partial M}(Ge)$, of an orbit $Ge \subseteq \Sigma^n(\rho)$ does not lie in $\Sigma^n(\rho)$. Define

$$\overset{\circ}{\Sigma}{}^n(\rho) := \{e \in \partial M \mid \text{cl}_{\partial M}(Ge) \subseteq \Sigma^n(\rho)\}.$$

There are important instances in which $\overset{\circ}{\Sigma}{}^n(\rho) = \Sigma^n(\rho)$ but when that is not the case $\overset{\circ}{\Sigma}{}^n(\rho)$ is easier to compute. In more detail:

- If ∂M is given the Tits distance topology then $\overset{\circ}{\Sigma}{}^n(\rho)$ is an open subset[3] of ∂M. In fact, $\{(\rho, e) \mid e \in \Sigma^n(\rho)\}$ is open in $\text{Hom}(G, \text{Isom}(M)) \times \partial M$. See Theorem F.
- The set $\{\rho \mid \Sigma^n(\rho) = \partial M\}$ is open in $\text{Hom}(G, \text{Isom}(M))$. See Theorem G. Note that, for such ρ, $\Sigma^n(\rho) = \overset{\circ}{\Sigma}{}^n(\rho)$.
- If M is almost geodesically complete then $\Sigma^n(\rho) = \partial M$ if and only if ρ is CC^{n-1}. See Theorem H.
- There is a description of $\overset{\circ}{\Sigma}{}^n(\rho)$ in terms of a dynamical condition on any free contractible G-CW-complex X over M; see Theorem E. This is the

[2] The space ∂M has two topologies: the usual compact "cone" topology and the "Tits distance" topology which is usually not compact but is complete and CAT(1). See Chapter 11 for details.

[3] The set $\overset{\circ}{\Sigma}{}^n(\rho)$ is not, in general, open in the cone topology; see the Remark following Theorem F in Chapter 10.

key to all our results on $\Sigma^n(\rho)$ and can be used to compute $\overset{\circ}{\Sigma}{}^n(\rho)$ in specific cases.

When M is Euclidean space $\mathbb{E}^m$ then ∂M is a sphere S^{m-1} and the two topologies on ∂M coincide; if the G-action ρ on $M = \mathbb{E}^m$ is by translations then the induced action on ∂M is trivial. In this situation our results were previously known: the two sets $\overset{\circ}{\Sigma}{}^n(\rho)$ and $\Sigma^n(\rho)$ coincide and have a description in terms of the *homotopical geometric invariant* $\Sigma^n(G)$ of the group G, on which a considerable literature exists. More specifically, if we take M to be the vector space $G/G' \otimes \mathbb{R}$ endowed with an inner product, and ρ to be the action of G on M given by left translation then $\Sigma^n(\rho)$ has a direct interpretation as $\Sigma^n(G)$. See Theorem I. $\Sigma^n(G)$ has been computed in many cases: see §§10.6 and 10.7(A). Explicit computation of $\Sigma^n(\rho)$ in non-Euclidean examples is still in its infancy, but we have recently solved the case of the natural action of $G = \mathrm{SL}_2(\mathbb{Z}[\frac{1}{m}])$, m a natural number, on the hyperbolic plane (§10.7(B)), and we compute $\overset{\circ}{\Sigma}{}^n(\rho)$ for all actions ρ of G on locally finite simplicial trees (§10.7(C)).

These ideas have evolved over the past twenty years. $\Sigma^1(G)$ was first introduced in the special context of metabelian groups G in [**BS 80**] as a tool to characterize finite presentability[4] of G. For arbitrary groups G, $\Sigma^1(G)$ was introduced in [**BNS 87**] and subsequently $\Sigma^n(G)$, $n \geq 2$, appeared in [**BRe 88**] and [**Re 88**], [**Re 89**]. Although these "geometric invariants" were defined for arbitrary groups G of type F_n they only referred to translation actions on Euclidean spaces, and so G was tacitly assumed to have infinite abelianization G/G'. The present work emerged when we tried to extend the scope of the theory to genuinely non-commutative actions. Not only the theorems but also the rather involved techniques concerning finitary maps and their sheaves have precursors in these earlier stages.

A full outline of this Part II is found in Chapter 10 below. The reader familiar with the self-contained essay on finitary sheaves and finitary maps which is Chapter 4 can read this second part with references back for some details and analogies.

[4]Precisely: a finitely generated metabelian group G is finitely presented if and only if for every point e on the sphere at infinity either $e \in \Sigma^1(G)$ or $-e \in \Sigma^1(G)$.

CHAPTER 10

Outline, Main Results and Examples

10.1. The boundary of a CAT(0)-space

Let (M, d) be a proper CAT(0) space. A *geodesic ray* in M is an isometric embedding $\gamma : [0, \infty) \to M$. Two geodesic rays γ, γ' are *asymptotic* if there is a constant $r \in \mathbb{R}$ such that $d(\gamma(t), \gamma'(t)) \leq r$ for all t. The set of all geodesic rays asymptotic to γ is called the *endpoint of* γ and denoted $e = \gamma(\infty)$. The collection of all endpoints of geodesic rays form the *boundary* ∂M of M. Since d is proper it is complete, so (see [**BrHa**] or [**Ho 97**]) for every pair $(a, e) \in M \times \partial M$ there is a unique geodesic ray $\gamma : [0, \infty) \to M$ with $\gamma(0) = a$ and $\gamma(\infty) = e$. Hence there is a natural bijection between ∂M and the set of all geodesic rays emanating from a base point, and ∂M acquires the compact-open topology of the latter via this bijection. This topology, which is independent of the choice of a, is compact and metrizable. It is called the *cone topology.* An action on M by isometries induces a topological action on ∂M. ∂M is assumed to carry the cone topology except when another interesting topology, the Tits distance topology (see §11.4), is explicitly mentioned.

Associated to each geodesic ray γ of M is its *Busemann function* $\beta_\gamma : M \to \mathbb{R}$ (see §11.2) and for each $s \in \mathbb{R}$ the associated *horoball* $HB_s(\gamma) := \beta_\gamma^{-1}([s, \infty))$. Horoballs "centered" at γ play a role analogous to that of balls centered at $a \in M$. Indeed, $HB_s(\gamma) = \mathrm{cl}_M(\bigcup\{B_{t-s}(\gamma(t)) | s < t\})$. Horoballs are contractible.

10.2. CC^{n-1} over end points

In parallel with §2.2, M is a proper CAT(0) space, G is a group, X^n an n-dimensional $(n-1)$-connected free G-CW-complex such that $G \backslash X^n$ is finite, $\rho : G \to \mathrm{Isom}(M)$ is an isometric action and $h : X \to M$ is a control function (i.e. a G-map). We pick a geodesic ray $\gamma : [0, \infty) \to M$ and its endpoint $e = \gamma(\infty)$, and we write $X_{(\gamma,s)}$ for the largest subcomplex of X lying in $h^{-1}(HB_s(\gamma))$. We say that *X is controlled $(n-1)$-connected (CC^{n-1}) in the direction γ* (with respect to ρ)[1] if for any horoball $HB_s(\gamma)$ and $-1 \leq p \leq n-1$ there exists $\lambda(s) \geq 0$ such that every map $f : S^p \to X_{(\gamma,s)}$ extends to a map $\tilde{f} : B^{p+1} \to X_{(\gamma, s-\lambda(s))}$ and $s - \lambda(s) \to \infty$ as $s \to \infty$. The number λ depends on the horoball $HB_s(\gamma)$ and is called a *lag.* In what follows, the lag will often turn out to be constant (i.e. independent of $HB_s(\gamma)$; dependent only on e), and then we will speak of a *constant lag.* When $p = -1$ this says that each $X_{(\gamma,s)}$ is non-empty.

The property "X is CC^{n-1} in the direction γ" is shown in §12.1 to be a property of the endpoint e rather than the ray γ. Moreover we have an Invariance Theorem (Theorem 12.1) as in Part I which shows that this property is also independent of

[1]A discussion of equivalent formulations of this definition can be found in the Appendices.

the choice of X and of h, i.e., it is a property of the action ρ. So, if X is CC^{n-1} in the direction γ we will say that ρ is CC^{n-1} *over* (or *in the direction*) $e = \gamma(\infty)$.

So far the two conditions CC^{n-1} over points $a \in M$, defined in §2.2, and endpoints $e \in \partial M$ look strictly analogous. But there is a striking difference in their behavior with respect to changing a or e: if ρ is CC^{n-1} over some $a \in M$ it is also CC^{n-1} over any other point of M. Not so for the property CC^{n-1} over $e \in \partial M$. Hence it becomes interesting to study the subset of ∂M

$$\Sigma^n(\rho) = \{e \mid \rho \text{ is } CC^{n-1} \text{ over } e\}.$$

EXAMPLE 10.1. Every homomorphism $G \to \mathbb{R}$ defines an action of G on $\mathbb{R}$ by translations (because of the canonical identification $\mathbb{R} \cong \text{Transl}(\mathbb{R})$) which is cocompact whenever the homomorphism is non-trivial. Let G be the Baumslag-Solitar group $\langle x, t \mid t^{-1}xt = x^2\rangle$. The homomorphism $x \mapsto 0$ and $t \mapsto 1$ defines such a cocompact action of G on $\mathbb{R}$. Now, $\partial\mathbb{R} = \{\pm\infty\}$. We claim that this action ρ is CC^∞ at $-\infty$ and is not CC^0 at $+\infty$, i.e. that $\Sigma^n(\rho) = \{-\infty\}$ for $n \geq 1$ while $\Sigma^0(\rho) = \partial\mathbb{R}$. To see this, first consider G as the ascending HNN extension $\mathbb{Z}*_\phi$ where $\phi : \mathbb{Z} \to \mathbb{Z}$ is multiplication by 2, and let T be the corresponding cocompact rooted[2] Bass-Serre G-tree. The tree T is the complete tree of valence 3. It is given the usual metric d, each edge having length 1. Our claim follows immediately from the existence of G-maps $T \times \mathbb{R} \xrightarrow{h_1} T \xrightarrow{h_2} \mathbb{R}$ where G acts freely, properly discontinuously and cocompactly on $T \times \mathbb{R}$, h_1 is projection on the T-factor, and h_2 extends to a map $\partial T \to \partial\mathbb{R}$ taking the root to $-\infty$ and all other end points of T to $+\infty$. (Simply take $T \times \mathbb{R}$ to be X, $\mathbb{R}$ to be M and $h_2 \circ h_1$ to be the control function h). It only remains to explain the constructions.

Ad h_2: Choose a base vertex v_0 for T. A rooted based G-tree always defines a homomorphism $\alpha : G \to \mathbb{R}$; to define $\alpha(g)$ find $u \in T$ such that there are (geodesic) segments in T from u to v_0 and from u to gv_0: then $\alpha(g) := d(gv_0, u) - d(v_0, u)$. Observe that in our case α is the homomorphism $x \mapsto 0$, $t \mapsto 1$. The corresponding map $h_2 : T \to \mathbb{R}$ which maps the vertex $v = gv_0$ to $\alpha(g)$ and is extended affinely to the edges of T is thus a G-map.

Ad h_1: Let U be the mapping torus of the map of degree $2 : S^1 \to S^1$, $z \mapsto z^2$. Then U is the presentation complex for our presentation of G, so G acts freely, properly discontinuously and cocompactly on the universal cover $\tilde{U}$. The proof that $\tilde{U}$ is homeomorphic to $T \times \mathbb{R}$ and the construction of the G-map h_1 (where $T \times \mathbb{R}$ is identified with $\tilde{U}$ as G-spaces) are given[3] in [**SW 79**, pp. 155-6, 166]. Alternatively, the reader who carefully considers the universal cover $\tilde{U}$ combinatorially will see it as $T \times \mathbb{R}$ with projection onto T being a G-map.

Following the tradition of [**BS 80**], [**BNS 87**], [**BRe 88**] and [**Re 88**], [**Re 89**] we call $\Sigma^n(\rho)$ the nth *(homotopical) geometric invariant of the action* ρ.

---

[2]"Rooted" means that G fixes an end point, a consequence of "ascending" which means that one of the subgroups in the HNN decomposition is the whole base group.

[3]Not all of this is explicit in those pages of [**SW 79**]. What is actually true (by the same construction and proof) is this **Theorem:** *Let $\phi : H \rightarrowtail H$ be a monomorphism and let Y be a connected complex with $\pi_1(Y) \cong H$. Assume ϕ is induced by a covering transformation $p : Y \to Y$ and let T be the Bass-Serre tree of $G := \langle H, t \mid t^{-1}ht = \phi(h), \forall\, h \in H\rangle$. If U is the presentation complex then $\tilde{U}$ is homeomorphic to $T \times \tilde{Y}$ and (identifying these) the projection $T \times \tilde{Y} \to T$ is a G-map.* This gives other examples which are CC^0 at $-\infty$ but not CC^0 at $+\infty$.

10.3. The dynamical subset

¿From the definition it is clear that if $e \in \partial M$ is in the geometric invariant $\Sigma^n(\rho)$, so is the whole G-orbit Ge. The subset of $\Sigma^n(\rho)$

$$\overset{\circ}{\Sigma}{}^n(\rho) := \{e \in \partial M | \mathrm{cl}_{\partial M}(Ge) \subseteq \Sigma^n(\rho)\}$$

deserves special attention. We call it the *dynamical subset* of $\Sigma^n(\rho)$ because the following theorem characterizes it in terms of a dynamical condition in the free contractible G-CW-complex X over M (with cocompact n-skeleton) of §10.2. In the spirit of §2.6 we call a cellular map $\phi : X^n \to X^n$ a *contraction towards* $e \in \partial M$ if there is a number $\varepsilon > 0$ with

$$\beta_\gamma h\phi(x) \geq \beta_\gamma h(x) + \varepsilon, \text{ for all } x \in X^n,$$

where $\gamma : [0, \infty) \to M$ is a geodesic ray with $\gamma(\infty) = e$. We have the following endpoint analogue of Theorem D:

THEOREM E. *If* $\rho : G \to \mathrm{Isom}(M)$ *is an action of a group* G *of type* F_n *on a proper* CAT(0) *space* M *by isometries then the dynamical subset* $\overset{\circ}{\Sigma}{}^n(\rho)$ *of* $\Sigma^n(\rho)$ *can be characterized as*

$$\overset{\circ}{\Sigma}{}^n(\rho) = \{e | X^n \textit{ admits a } G\textit{-finitary contraction towards } e\}.$$

Note that $\overset{\circ}{\Sigma}{}^n(\rho) = \partial M$ if and only if $\Sigma^n(\rho) = \partial M$. The example of a Fuchsian group acting on the hyperbolic plane (see §10.7(B)) shows that $\Sigma^n(\rho)$ and $\overset{\circ}{\Sigma}{}^n(\rho)$ are in general different but they contain the same closed G-invariant subsets of ∂M.

10.4. Openness results

The "angular distance" between two points e and e' of ∂M is the supremum over points $a \in M$ of the angle between the geodesic rays γ and γ' representing e and e' where $\gamma(0) = \gamma'(0) = a$. This is a metric on ∂M and the corresponding length metric is called the "Tits distance", denoted $Td(e, e')$; see §11.4 for full definitions. In general, the topology on ∂M given by Td is finer than the cone topology: i.e. Id: $(\partial M, Td) \to (\partial M$, cone topology) is continuous. The space $(\partial M, Td)$ is a complete CAT(1) metric space [**BrHa**, III 3.17]. Two extremes are represented by $M = \mathbb{E}^k$ where the two topologies agree giving S^{k-1}, and $M = \mathbb{H}^k$ where $(\partial M, Td)$ is discrete while $(\partial M$, cone topology) is S^{k-1}.

We have openness theorems involving both topologies on ∂M. The easier of these involves Tits distance. Recall from §7.1 that the function spaces Isom(M) and Hom(G,Isom(M)) are given the compact-open topology. With this understanding, the condition "$e \in \overset{\circ}{\Sigma}{}^n(\rho)$" is open with respect to both ρ and e:

THEOREM F. $\{(\rho, e) \mid e \in \overset{\circ}{\Sigma}{}^n(\rho)\}$ *is open in* $\mathrm{Hom}(G, \mathrm{Isom}(M)) \times \partial M$ *when* ∂M *carries the Tits distance topology. In particular, for fixed* ρ, $\overset{\circ}{\Sigma}{}^n(\rho)$ *is open in* $(\partial M, Td)$.

REMARK. In general, neither $\Sigma^n(\rho)$ nor $\overset{\circ}{\Sigma}{}^n(\rho)$ is open in $(\partial M$, cone topology). If h is a non-trivial element of the free group of rank 2 and ρ is the action of $G := \langle h \rangle$ on the Cayley tree T by covering translations, then $\overset{\circ}{\Sigma}{}^n(\rho)$ consists of the two endpoints of the translation axis of h; this is not open in ∂T. For an example where $\Sigma^n(\rho)$ is not open see §10.7(B): that example also shows that $\overset{\circ}{\Sigma}{}^n(\rho)$ is not always the interior of $\Sigma^n(\rho)$ in $(\partial M, Td)$. We do not know if $\Sigma^n(\rho)$ is always open in $(\partial M, Td)$. In the "classical" case of [**BS 80**], [**BNS 87**], [**BRe 88**] and [**Re 88**], [**Re 89**] (discussed in §10.7(A)) $\Sigma^n(\rho) = \overset{\circ}{\Sigma}{}^n(\rho)$ and is open in S^{k-1}.

For a different and deeper openness theorem we need ∂M compact, i.e. the cone topology:

THEOREM G. *If E is a closed subset of $(\partial M$, cone topology$)$ then $\{\rho \mid E \subseteq \Sigma^n(\rho)\}$ is an open subset of* Hom$(G,$Isom$(M,E))$. *In particular, $\{\rho \mid \Sigma^n(\rho) = \partial M\}$ is open in* Hom$(G,$Isom$(M))$.

Here, Isom(M,E) is the space of isometries of M which leave E invariant.

10.5. Endpoints versus points in M

We come to a theorem which relates the CC^{n-1} property over all the endpoints $e \in \partial M$ to the CC^{n-1} property over a point $a \in M$. This is the link between the present paper and Part I; it requires a mild assumption on M. The (proper) CAT(0) space (M,d) is *geodesically complete* if every geodesic segment $[0,t] \to M$ can be extended to a geodesic ray $[0,\infty) \to M$. We say the CAT(0) space (M,d) is *almost geodesically complete* if there is a number $\mu \geq 0$ such that for any two points $a, b \in M$ there is a geodesic ray γ with $\gamma(0) = a$ such that $\gamma([0,\infty))$ meets $B_\mu(b)$. A recent Theorem of P. Ontaneda shows that this property is often guaranteed in cases of interest[4].

THEOREM H. **(Boundary Criterion)** *If $\rho : G \to$ Isom(M) is an isometric action on a proper and almost geodesically complete* CAT(0)*-space M then the following are equivalent*

(i) $\Sigma^n(\rho) = \partial M$;
(ii) ρ *is* CC^{n-1}.

We do not know whether the assumption that M be almost geodesically complete is necessary in Theorem H. It is not needed in the case $n = 0$, where the Theorem says that $\Sigma^0(\rho) = \partial M$ if and only if ρ is cocompact. It is an open problem as to whether there exists a non-compact CAT(0) space M which is not almost geodesically complete but admits a cocompact group action by isometries.[5]

The proofs of Theorems E, F, G and H are completed in Chapter 15.

---

[4]**Theorem.** [**On**] *Let M be a non-compact proper* CAT(0)*-space such that* Isom(M) *acts cocompactly. A sufficient condition for almost geodesic completeness is that the cohomology with compact supports $H_c^*(M)$ be non-trivial. This condition is satisfied whenever some subgroup of* Isom(M) *acts cocompactly with discrete orbits.*
Recently, D. Farley has shown that this condition is also satisfied if M (as above) is an M_κ-complex with finite shapes (see [**BrHa**] for the relevant definitions).

[5]The conclusion of Theorem B ("cocompact and CC^{n-1}" is an open condition) follows from Theorems G and H. This alternative proof of Theorem B has the disadvantage that it requires the (perhaps redundant) hypothesis that M be almost geodesically complete. It has the advantage of being technically simpler, in that ∂M (with the cone topology) plays the role previously played by the sphere of event radius R (§5.3): the compactness of this sphere is replaced by the compactness of ∂M. The reader who carefully compares these two proofs of Theorem B will see that this alternative proof is easier, avoiding most of the content of §§6.3, 7.3 and 7.4.

10.6. Fixed points and the BNSR-geometric invariant

At this point we should relate our theory to the previous literature. Let $S(G)$ be the sphere of non-zero (additive) characters on G modulo positive scalar multiplication: when $\chi \in \text{Hom}(G, \mathbb{R})$ is a non-zero character we denote by $[\chi]$ the point of $S(G)$ represented by χ. Identify $\mathbb{R}$ with the group of translations of the Euclidean line and interpret χ as a translation action of G on $\mathbb{R}$. The preferred endpoint of $\mathbb{R}$ is ∞. Define

$$\Sigma^n(G) := \{[\chi] \in S(G) \mid \chi \text{ is } CC^{n-1} \text{ over } \infty\}.$$

This is the *homotopical geometric invariant* of [**BRe 88**] and [**Re 88**], and coincides with the invariant $\Sigma_{G'}$ of [**BNS 87**] when $n = 1$.

Now we consider a CAT(0) space M with a specified end point $e \in \partial M$. We denote by $\text{Isom}(M, e)$ the group of isometries which fix e. If $f : M \to M$ is an isometry fixing e then f permutes the horoballs centered at e, and so $f(HB_r(\gamma)) = HB_{r'}(\gamma)$ for γ a geodesic ray with $\gamma(\infty) = e$ and $r \in \mathbb{R}$. It is not difficult to observe that the map $r \mapsto r'$ is a translation of $\mathbb{R}$ (see Proposition 11.3) so that we have a canonical map

$$\kappa : \text{Isom}(M, e) \to \mathbb{R} = \text{Isom}(\mathbb{R}, \infty).$$

Thus every G-action $\rho : G \to \text{Isom}(M, e)$ induces a translation action $\kappa\rho$ on $\mathbb{R}$ – in other words, an additive homomorphism $\chi_{\rho,e} : G \to \mathbb{R}$. Moreover, the Busemann function $\beta_\gamma : M \to \mathbb{R}$ is compatible with the actions ρ and $\kappa\rho$ on M and $\mathbb{R}$, so if $h : X \to M$ is the control function chosen in §2.2 with respect to ρ we can choose the composition $h_e = \beta_\gamma h : X \to \mathbb{R}$ as the control function for $\kappa\rho$. It is then immediate from the definitions in §10.2 that ρ is CC^{n-1} over e if and only if $\kappa\rho$ is CC^{n-1} over ∞. This proves

THEOREM I. *Let $\rho : G \to \text{Isom}(M, e)$ be an action fixing the point $e \in \partial M$. Then $e \in \Sigma^n(\rho)$ (or, equivalently, $e \in \mathring{\Sigma}^n(\rho)$) if and only if $[\kappa\rho] \in \Sigma^n(G)$.* □

A more intimate relationship between $\Sigma^n(\rho)$ and $\Sigma^n(G)$ can be observed if $M = \mathbb{E}^k$ is Euclidean k-space and $\rho : G \to \text{Transl}(\mathbb{E}^k)$ is an action on $\mathbb{E}^k$ by translations. Then $\partial\mathbb{E}^k$ is a $(k-1)$-sphere and is pointwise fixed by the G-action, so that $e \mapsto [\chi_{\rho,e}]$ defines a map[6]

$$\mu : \partial\mathbb{E}^k = S^{k-1} \to S(G) \cup \{0\}.$$

Let N be the linear span of the orbit of 0, and let N' be its orthogonal complement in $\mathbb{E}^k$. Then $\partial\mathbb{E}^k = \partial N * \partial N'$ is the join of great subspheres in $\partial\mathbb{E}^k$. The translation action ρ of G on $\mathbb{E}^k$ restricts faithfully to a translation action ρ_N on N, and $\Sigma^n(\rho) = \Sigma^n(\rho_N) * \partial N' - \emptyset * \partial N'$. The map μ respects the join operation, is injective on ∂N and maps $\partial N'$ to 0, so it maps $\Sigma^n(\rho_N)$ bijectively onto the intersection of $\Sigma^n(G)$ with the great subsphere $\{[\chi] \mid \chi(\ker \rho) = 0\}$ of $S(G)$. Thus the invariant $\Sigma^n(\rho)$ is determined by $\Sigma^n(G)$. In the special case where ρ is cocompact and G acts properly discontinuously, μ maps $\Sigma^n(\rho)$ onto $\Sigma^n(G)$ homeomorphically.

10.7. Examples

A) **Actions by Euclidean translations.** Let G be a group of type $F_n, n \geq 1$, and $M = G/G' \bigotimes \mathbb{R}$ the real G-vector space endowed with a Euclidean metric. The action $\rho : G \to \text{Transl}(M)$ is induced by the left regular action of G on G/G'. ∂M

[6] $S(G) \cup \{0\}$ is a quotient space of $\text{Hom}(G, \mathbb{R})$, $\{0\}$ is dense and μ is continuous.

is a $(k-1)$-sphere ($k = \dim_{\mathbb{R}}(G/G' \bigotimes \mathbb{R})$) which is pointwise fixed by ρ. Thus, by §10.6, we have

$$\overset{\circ}{\Sigma}{}^n(\rho) = \Sigma^n(\rho) = \Sigma^n(G) \subseteq S(G).$$

The invariant $\Sigma^n(G)$ has been computed in many specific cases. We outline two situations which demonstrate the subtle behavior of $\Sigma^n(G)$.

A1) **Right angled Artin groups.** Let Γ be a graph. The group $G(\Gamma)$ is given by the presentation whose generators are the vertices of Γ subject to the relations that adjacent vertices commute. Every vertex $v \in \text{ver}(\Gamma)$ gives rise to the *coordinate hemispheres* $\pm H_v \subset S(G(\Gamma))$ where

$$H_v := \{[\chi] | \chi \in \text{Hom}(G(\Gamma), \mathbb{R}), \chi(v) > 0\}.$$

Bestvina and Brady [**BeBr 97**] have shown that the diagonal character $\chi : G(\Gamma) \to \mathbb{R}$, defined by $\chi(v) = 1$ for all $v \in \text{ver}(\Gamma)$, lies in $\Sigma^n(G)$ if and only if the flag complex[7] of Γ is $(n-1)$-connected. Meier, Meinert, and VanWyk [**MMV 98**] have extended this to a complete description of $\Sigma^n(G(\Gamma))$; it is a finite union of finite intersections of coordinate hemispheres. For a simple geometric approach to these results see [**BuGo 99**].

A2) **Metabelian groups.** A group G is said to be *metabelian* if its commutator subgroup G' is abelian. For finitely generated metabelian groups G the invariant $\Sigma^1(G)$ has a handy description in terms of valuations on fields [**BS 81**] which can be used for calculations. In [**BGr 84**] it was used to prove that $\Sigma^1(G)$ is a rational polyhedral subset of $S(G)$, i.e., a finite union of finite intersections of open hemispheres with rational coordinate ratios for the spherical centers. It has been conjectured that in the metabelian case $\Sigma^n(G)$ is determined by $\Sigma^1(G)$; more precisely, writing $\Sigma^n(G)^c$ for the complement of $\Sigma^n(G)$ in $S(G)$, the Σ^n-*Conjecture* reads[8]: G being of type F_n,

$$\begin{aligned}\Sigma^n(G)^c = &\{[\chi] \in S(G) \mid \chi = \chi_1 + \cdots + \chi_k \\ &\text{with } k \leq n \ \text{ and each } \ [\chi_i] \in \Sigma^1(G)^c\}\end{aligned}$$

This conjecture has been verified when the commutator subgroup G' is virtually torsion-free of finite rank by H. Meinert [**Me 96**] and when G' is torsion and of Krull dimension 1 as a $\mathbb{Z}G/G'$-module by D. Kochloukova [**Ko 96**]. Moreover, Kochloukova has also proved it for general split[9] metabelian groups and $n = 2$. We will see examples in C) below.

B) **$\mathbf{SL_2}(\mathbb{R})$-action on the hyperbolic plane.** Let G be a discrete subgroup of $\text{SL}_2(\mathbb{R})$ (a Fuchsian group) and let $\rho : G \to \text{Isom}(\mathbb{H}^2)$ be the natural action of G on the hyperbolic plane $\mathbb{H}^2$. If this action has a fundamental domain of finite area, then the complement of $\Sigma^0(\rho)$ in $\partial\mathbb{H}^2$ is precisely the set of all parabolic fixed

[7]The flag complex of a graph is the simplicial complex whose n-simplices are the $(n+1)$-element sets of pairwise adjacent vertices

[8]The Σ^n-Conjecture for metabelian groups G came up in 1988. It was based on the older FP_n-*Conjecture* (A finitely generated metabelian group G is of type FP_n if and only if every n-point subset of $\Sigma^1(G)^c$ is contained in an open hemisphere; see [**B 81**].). Indeed, one can observe that if $G \rightarrowtail \tilde{G} \twoheadrightarrow \mathbb{Z}$ is a short exact sequence then the FP_n-Conjecture for $\tilde{G}$ implies the Σ^n-Conjecture for the rational points in $\Sigma^n(G)$ up to antipodality.

[9]i.e. the commutator subgroup $G' \trianglelefteq G$ has a complement in G.

points of G; or, equivalently, the orbits of the points $e \in \partial\mathbb{H}^2$ which lie on the boundary of a fundamental polygon (see [**BG 98**]). It is not difficult to observe that, in fact, $\Sigma^n(\rho) = \Sigma^0(\rho)$ for all n. Since the action of G on $\partial\mathbb{H}^2$ has only dense orbits we have $\mathring{\Sigma}^n(\rho) = \emptyset$ for all n.

A special case of a Fuchsian group is $\mathrm{SL}_2(\mathbb{Z})$. If we use the upper half plane model for $\mathbb{H}^2$, the boundary is $\partial\mathbb{H}^2 = \mathbb{R} \cup \{\infty\}$ and $\Sigma^0(\rho)^c$ consists of the single orbit $\mathrm{SL}_2(\mathbb{Z})\infty = \mathbb{Q} \cup \{\infty\} =: \overline{\mathbb{Q}}$.

Let $S = \{p_1, \dots, p_s\}$ be a finite set of s (different) prime numbers in $\mathbb{N}$ and let $\mathbb{Z}_S$ be the subring of $\mathbb{Q}$ generated by $(p_1 p_2 \dots p_s)^{-1}$. We let $\rho : \mathrm{SL}_2(\mathbb{Z}_S) \to \mathrm{Isom}(\mathbb{H}^2)$ be the natural action of $\mathrm{SL}_2(\mathbb{Z}_S)$ on the upper half plane by Moebius transformations. This action has dense orbits so, by Theorem H, $\mathring{\Sigma}^n(\rho)$ is empty or all of $\partial\mathbb{H}^2$, and ρ is CC^{n-1} over $\infty \in \partial\mathbb{H}^2$ if and only if ρ is CC^{n-1}. For this ρ we prove in [**BG**] the following:

THEOREM.

$$\Sigma^n(\rho)^c = \begin{cases} \emptyset & \text{if } n < s \\ \overline{\mathbb{Q}} & \text{if } n \geq s \end{cases}$$

In particular

$$\mathring{\Sigma}^n(\rho)^c = \begin{cases} \emptyset & \text{if } n < s \\ \partial\mathbb{H}^2 & \text{if } n \geq s \end{cases}$$

Part of the interest of this theorem has been discussed in §2.8, but it is now given in a sharper form. Eventually one would hope for a similar statement for $\mathrm{SL}_2(O_S)$ where O_S is the ring of S-integers in an algebraic number field.

C) **Tree actions**. Let T be an infinite locally finite tree and let $\rho : G \to \mathrm{Isom}(T)$ be a cocompact action of G by simplicial automorphisms. Then, by Bass-Serre theory, G is the fundamental group of a finite graph of groups $(\Gamma, \mathcal{G})$, where $\Gamma = G\backslash T$ and $\mathcal{G}$ is the system of edge and vertex stabilizers along a fundamental transversal of T. The edge stabilizers are of finite index in the vertex stabilizers since T is locally finite. Following [**B 98**] we define the *finiteness length* of G [resp. $\mathcal{G}$] to be $\mathrm{fl}\, G := \sup\{k \mid G \text{ is of type } F_k\}$ [resp. $\mathrm{fl}\, \mathcal{G} := \inf\{\mathrm{fl}\, H \mid H \in \mathcal{G}\}$], and the *connectivity length* of a character $\chi : G \to \mathbb{R}$ to be $\mathrm{cl}(\chi) := \sup\{k \mid k \leq \mathrm{fl}\, G$ and $[\chi] \in \Sigma^k(G)\}$. In this case, $\mathrm{fl}\, \mathcal{G} = \mathrm{fl}\, H$ for any $H \in \mathcal{G}$.

We begin by noting three elementary facts. First, T is almost geodesically complete so we may apply Theorem H. Secondly, if the fixed point set $(\partial T)^G$ is a proper subset of ∂T then it is either empty or a singleton; for if there are two singleton orbits then ∂T consists of just those points. Thirdly, any orbit consisting of more than one point is dense, so that its closure is ∂T. It follows that if S is the union of closures of orbits—and $\mathring{\Sigma}^n(\rho)$ is such a set—then $S = \partial T$ or is a singleton or is empty. It is well known that $\mathrm{fl}\, \mathcal{G} \leq \mathrm{fl}\, G$, so *if* $(\partial T)^G = \emptyset$ *then*[10]

$$\mathring{\Sigma}^n(\rho) = \begin{cases} \partial T & \text{if } 0 \leq n \leq \mathrm{fl}\, \mathcal{G} \\ \emptyset & \text{if } \mathrm{fl}\, \mathcal{G} < n \leq \mathrm{fl}\, G \end{cases} \tag{10.1}$$

There remains the case when the G-tree T has exactly one fixed end e. For such a tree we have the associated non-zero character $\chi_{\rho,e} : G \to \mathbb{R}$ of §10.6 measuring the shift towards e. Thus if $n \leq \mathrm{fl}\, \mathcal{G}$, $\Sigma^n(\rho)$ is defined, and by Theorems A and H,

[10]When $\mathrm{fl}\, G$ or $\mathrm{fl}\, \mathcal{G} = \infty$, this is to be understood as meaning "for all n".

$\Sigma^n(\rho) = \partial T$. In particular, $e \in \Sigma^n(\rho)$ and therefore, by Theorem I, $[\chi_{\rho,e}] \in \Sigma^n(G)$, implying $n \leq \mathrm{cl}(\chi_{\rho,e})$. In summary: $\mathrm{fl}\ \mathcal{G} \leq \mathrm{cl}(\chi_{\rho,e}) \leq \mathrm{fl}\ G$ and

$$\mathring{\Sigma}^n(\rho) = \begin{cases} \partial T & \text{if } 0 \leq n \leq \mathrm{fl}\ \mathcal{G} \\ \{e\} & \text{if } \mathrm{fl}\ \mathcal{G} < n \leq \mathrm{cl}(\chi_{\rho,e}) \\ \emptyset & \text{if } \mathrm{cl}(\chi_{\rho,e}) < n \leq \mathrm{fl}\ G \end{cases} \tag{10.2}$$

By a *non-trivial rooted G-tree* we mean a (simplicial) G-tree with no fixed point and with a unique fixed end e (neither cocompactness nor local finiteness are assumed here). Every such G-tree $\rho : G \to \mathrm{Isom}(T)$ has an associated non-zero character $\chi_{\rho,e} : G \to \mathbb{R}$, defining an associated point $[\chi_{\rho,e}] \in S(G)$. When G is finitely generated then for any non-trivial locally finite rooted G-tree there is a (cocompact) Bass-Serre tree of an ascending HNN extension of G, i.e.,

$$G = \langle B, t \mid tbt^{-1} = \theta(b), b \in B \rangle, \tag{10.3}$$

where $\theta : B \rightarrowtail B$ is an injective endomorphism and $\theta(B)$ has finite index ≥ 2 in B, so that the two trees have the same associated point of $S(G)$; i.e. one associated character is a positive multiple of the other[11].

We will examine the parameters $\mathrm{fl}\ \mathcal{G}$, $\mathrm{cl}(\chi_{\rho,e})$ and $\mathrm{fl}\ G$ occurring in (10.2) when G belongs to a special class of finitely generated groups. For this, we define (for any character $\chi : G \to \mathbb{R}$ on any finitely generated group)

$$m(\chi) := \sup\{k \mid \chi \neq \chi_1 + \cdots + \chi_k \ \text{ with each } \ [\chi_i] \in \Sigma^1(G)^c - \{[\chi]\}\}.$$

Our special class is the class of finitely generated *MFPR groups*, i.e. metabelian groups of finite Prüfer rank. Recall that this means: the commutator subgroup G' is abelian with finite torsion and finite torsion free rank ($\dim_{\mathbb{Q}}(G' \otimes \mathbb{Q}) < \infty$). For such groups there is no difference between type F_n and type FP_n ([**BS 80**, Theorem 5.4]). Both the Σ^n-Conjecture and the FP_n-Conjecture (see §10.6) are known to hold for MFPR groups ([**Å 86**], [**Me 96**]). Denoting the zero character by 0_G and an arbitrary character by χ, these theorems can be restated in terms of the function m:

$$\begin{cases} \mathrm{fl}\ G = m(0_G) \\ \mathrm{cl}(\chi) = \min\{m(\chi), m(0_G)\}. \end{cases} \tag{10.4}$$

If G splits as in (10.3), the character $\chi_{\rho,e}$ of the corresponding tree decomposition is given by $\chi_{\rho,e}(B) = 0$ and $\chi_{\rho,e}(t) = -1$. According to [**Me 96**, Theorem 2.6], *provided B is finitely generated,*

$$\mathrm{fl}\ B = \min\{m(\chi_{\rho,e}), m(-\chi_{\rho,e}), m(0_G)\}. \tag{10.5}$$

The requirement that B be finitely generated is not a serious restriction if the MFPR group G is finitely presented. In that case, for any non-trivial rooted G-tree corresponding to a decomposition (10.3) there is another such, having the same associated point of $S(G)$, in which B is finitely generated [**BS 78**, Theorem A][12].

[11]More precisely, there is a G-invariant "superdivision" T_1 of a G-invariant subtree T' of T (i.e. T' is a subdivision of T_1) such that $G \backslash T_1$ consists of one vertex and one edge. This $T' := \cup\{gA(h) \mid g \in G\}$ where $A(h)$ is the translation axis of a hyperbolic element $h \in G$. There is always a hyperbolic element under these hypotheses.

[12]In fact there is a map of G-trees from the latter onto the former.

Summarizing: in this nice case (G a finitely presented MFPR group and B finitely generated) (10.2) can be rewritten:

$$\overset{\circ}{\Sigma}{}^n(\rho) = \begin{cases} \partial T & \text{if } 0 \leq n \leq \min\{m(\chi_{\rho,e}), m(-\chi_{\rho,e}), m(0_G)\} \\ \{e\} & \text{if } \min\{m(\chi_{\rho,e}), m(-\chi_{\rho,e}), m(0_G)\} < n \\ & \quad \leq \min\{m(\chi_{\rho,e}), m(0_G)\} \\ \emptyset & \text{if } \min\{m(\chi_{\rho,e}, m(0_G)\} < n \leq m(0_G). \end{cases} \tag{10.6}$$

Since the definition of $m(\chi)$ involves geometric features of the whole set $\Sigma^1(G)^c$, it is only a slight exaggeration to conclude from (10.6) that one must know $\Sigma^1(G)^c$ in order to know the $\overset{\circ}{\Sigma}{}^n$-properties of ρ.

What can $\Sigma^1(G)^c$ be? If G is a finitely generated MFPR group, then a precise answer can be given: $\Sigma^1(G)^c$ is a finite set of rational points in $S(G)$; moreover, given k and any finite rational subset A of S^k there is a finitely generated MFPR group G with $S(G) = S^k$ and $\Sigma^1(G)^c = A$; and in order to ensure that G is finitely presented, one simply chooses A so that it does not contain diametrically opposite points. See [**BS 81**, Example 2.6].

At this point we recall a theorem of K.S. Brown [**Br 87$_{\mathbf{II}}$**] which implies[13] that for any finitely generated group G the image of the map $\rho \mapsto [\chi_{\rho,e}]$ from the set of all non-trivial rooted G-trees into $S(G)$ consists of all the rational points of $\Sigma^1(G)^c$. It follows from our discussion that if we only consider G-trees as in (10.3) with B finitely generated (call these "special") then, provided G is finitely presented, (i) there are enough special G-trees to map onto the rational points of $\Sigma^1(G)^c$; (ii) those mapping to the same point have the same $\overset{\circ}{\Sigma}{}^n$-properties, and (iii) the $\overset{\circ}{\Sigma}{}^n$-properties of the various special G-trees are interdependent in a manner dictated by the location of the finite set $\Sigma^1(G)^c$ in $S(G)$.

A final remark: Among the choices of finitely generated MFPR groups G and $[\chi] \in \Sigma^1(G)^c$ we always have $m(\chi) = m(0_G)$ or $m(0_G) - 1$ (and both can occur), but we can make $m(\chi)$ and $m(\chi) - m(-\chi)$ as large as we like by choosing k and A, above, suitably. Thus, in (10.1) we can achieve $\overset{\circ}{\Sigma}{}^n(\rho) = \emptyset$ only when $n = \mathrm{fl}\ G < \infty$, while we can achieve $\overset{\circ}{\Sigma}{}^n(\rho) = \partial T$ and $\overset{\circ}{\Sigma}{}^n(\rho) = \{e\}$ in arbitrarily large ranges of n. We do not know if $\overset{\circ}{\Sigma}{}^n(\rho) = \emptyset$ can occur in a larger range when G is not an MFPR group.

[13]Brown's theorem asserts that $\Sigma^1(G)^c$ consists of all $[-\chi_{\rho,e}]$ such that ρ is a G-action on an $\mathbb{R}$-tree with no fixed point and a unique fixed end. For the rational points of $S(G)$, simplicial (but not necessarily locally finite) trees suffice.

CHAPTER 11

Further Technicalities on CAT(0) spaces

11.1. More on ∂M

In order to relate the topology of the CAT(0) space M to that of its boundary ∂M we use the following notation[1]: A *generalized geodesic ray* $\gamma : [0,\infty) \to M$ is either a geodesic ray in the sense of §10.1 or a geodesic segment $[0,\mu] \to M$ extended by $\gamma([\mu,\infty)) = \gamma(\mu)$; it *starts* at $\gamma(0)$ and *ends* at $\gamma(\infty) := \gamma(\mu)$.

Let R be the function space consisting of all generalized geodesic rays $\gamma : [0,\infty) \to M$ with the compact open topology. (i.e., uniform convergence on compact sets). Let $\hat{M}$ denote the disjoint union $M \cup \partial M$, and $\varepsilon : R \twoheadrightarrow \hat{M}$ the endpoint map $\varepsilon(\gamma) := \gamma(\infty)$. For each $a \in M$ the map ε has a continuous canonical section $\sigma_a : \hat{M} \rightarrowtail R$, where $\sigma_a(e)$ is the unique generalized geodesic ray starting at $a \in M$ and ending at $e \in \hat{M}$. This shows that $\hat{M}$ with the quotient topology induced by ε is homeomorphic to the subspace $R_a := \sigma_a(\hat{M}) = \{\gamma \in R | \gamma(0) = a\}$ of R. Restricting σ_a to the subspace $M \subset \hat{M}$ shows that the topology of M inherited from $\hat{M}$ is the topology given by the metric on M; and ∂M is homeomorphic to the subspace $\partial R_a := \sigma_a(\partial M) \subset R$ consisting of all geodesic rays emanating from a.

PROPOSITION 11.1. *$\hat{M}$ is a compact metrizable space containing M as a subspace, $\partial M = \hat{M} - M$ is a nowhere dense subset and for every open set $U \subset \hat{M}$, the inclusion $U - \partial M \rightarrowtail U$ is a homotopy equivalence.* □

When a group G acts on M by isometries there is an obvious extension to an action of G on $\hat{M}$ by homeomorphisms: If $e \in \partial M$ is the endpoint of the geodesic ray $\gamma : [0,\infty) \to M$, and $g \in G$, then $g\gamma$ is a geodesic ray and we put $ge := (g\gamma)(\infty)$. The stabilizer of $e \in \partial M$ in G is denoted G_e.

11.2. Review of Busemann functions and horoballs

Associated to a generalized geodesic ray $\gamma : [0,\infty) \to M$ is its Busemann function $\beta_\gamma : M \to \mathbb{R}$ defined by

$$\beta_\gamma(b) = \lim_{t\to\infty} (d(\gamma(0), \gamma(t)) - d(b, \gamma(t)), \qquad b \in M.$$

In the case when $\gamma(\infty) \in M$ and μ is the smallest non-negative number with $\gamma(\mu) = \gamma(\infty)$ we find $\beta_\gamma(b) = \mu - d(b, \gamma(\mu))$. In the case when $\gamma(\infty) \in \partial M$ one shows, by the triangle inequality, that the map $t \mapsto t - d(b, \gamma(t))$ is monotone increasing and bounded above by $d(\gamma(0), b)$, so $\beta_\gamma(b)$ is well defined.

PROPOSITION 11.2. (a) *If γ and γ' are generalized geodesic rays with the same endpoint then $\beta_\gamma(\cdot) - \beta_{\gamma'}(\cdot)$ is constant. In particular the difference $\beta_\gamma(a) - \beta_\gamma(b)$ depends only on a, b, and $e = \gamma(\infty)$.*

[1]For proofs of the assertions about CAT(0) spaces in this section see [**BrHa**].

(b) *The function* $\beta : R \times M \to \mathbb{R}$, $\beta(\gamma, b) = \beta_\gamma(b)$ *is continuous.* $\square$

If γ is a generalized ray and $s \in \mathbb{R}$, then the (closed) *horoball* $HB_s(\gamma)$ is $\beta_\gamma^{-1}([r, \infty))$. When $s \geq 0$ this has a more geometric interpretation, namely $HB_s(\gamma) = \mathrm{cl}_M(\cup\{B_{t-s}(\gamma(t)) | s < t\})$. If the ray $\gamma : [0, \infty) \to M$ is degenerate with $e = \gamma(\infty) \in M$ and $d(\gamma(0), e) = \mu$ then $HB_s(\gamma)$ is precisely $B_{\mu-s}(e)$.

11.3. G-actions and Busemann functions

Let G be a group acting on M by isometries, and let $\gamma : [0, \infty) \to M$ be a geodesic ray. For $g \in G$ and $a \in M$ we observe that $\beta_\gamma(ga) - \beta_\gamma(a)$ depends only on g, a and $e = \gamma(\infty)$. So putting $\psi_e(g, a) := \beta_\gamma(ga) - \beta_\gamma(a)$ defines a map $\psi_e : G \times M \to \mathbb{R}$. This map satisfies the equation

$$\psi_e(gh, a) = \psi_e(g, ha) + \psi_e(h, a), \qquad g, h \in G. \tag{11.1}$$

To see how $\psi_e(g, a)$ depends on a we compute

$$\begin{aligned} \psi_e(g, a) - \psi_e(g, b) &= \beta_\gamma(ga) - \beta_\gamma(a) - \beta_\gamma(gb) + \beta_\gamma(b) \\ &= (\beta_{g^{-1}\gamma}(a) - \beta_\gamma(a)) - (\beta_{g^{-1}\gamma}(b) - \beta_\gamma(b)), \end{aligned}$$

where we have used the obvious fact that $\beta_{gx}(ga) = \beta_x(a)$. Now suppose g lies in the stabilizer G_e of e. Then $g^{-1}\gamma$ and γ have the same endpoint and since $\beta_{g^{-1}\gamma}(\cdot) - \beta_\gamma(\cdot)$ is constant we find $\psi_e(g, a) = \psi_e(g, b)$ for all $g \in G_e$, $a, b \in M$. By (11.1) we get:

PROPOSITION 11.3. *The function* $\chi_e : G_e \to \mathbb{R}$ *defined by* $\chi_e(g) = \beta_\gamma(ga) - \beta_\gamma)$ *is an additive homomorphism. In other words,* $g \cdot r := r + \chi_e(g)$ *defines a translation action of* G_e *on* $\mathbb{R}$ *which makes the Busemann function* $\beta_\gamma : M \to \mathbb{R}$ *a* G_e*-map.* $\square$

11.4. The Tits distance

We need some preliminary definitions. If Δ is a geodesic triangle in M with vertices a, b and c, and if Δ' is the comparison triangle in the plane, the *angle at* a in Δ is the Euclidean angle at a in Δ'; compare §2.1. The *angle* between geodesic rays γ and γ' starting at the same point $a \in M$ is $\angle_a(\gamma, \gamma') := \lim_{\epsilon \to 0} \sup_{t,t' < \epsilon} \{$ the angle at $\gamma(0)$ with vertices $\gamma(t)$ and $\gamma'(t')\}$. The *angular distance* between points e and e' of ∂M is $\angle(e, e') := \sup_{a \in M}\{\angle_a(\gamma, \gamma')\}$ where γ and γ' (as above) represent e and e' respectively. This is a metric called the *angular metric* on ∂M: it is bounded above by π. The *Tits distance*, $Td(e, e')$ between $e, e' \in M$ is the inf of lengths of rectifiable (with respect to the angular metric) paths[2] in ∂M. These two metrics are complete and are locally isometric to one another. Hence they define the same topology on ∂M. The identity map $(\partial M, Td) \to (\partial M$, cone topology) is continuous. These metrics give rich structure to ∂M; in particular $(\partial M, Td)$ is a complete CAT(1) space [**BrHa**, III 3.17] where "CAT(1)" is defined in the same way as CAT(0) but using comparison triangles in S^2 rather than in the plane.

[2]If there are no such paths the Tits distance is defined to be ∞; this extension of the usual notion of "metric" causes no problems.

CHAPTER 12

CC^{n-1} over Endpoints

12.1. Invariance Theorem

As in §3.1 M is a proper CAT(0) space, X is a free left G-complex with $G\backslash X^n$ finite and $\rho : G \to \text{Isom}(M)$ is a left action of G on M by isometries. By Proposition 3.1 we can choose a G-equivariant control function $h : X \to M$.

A *decreasing filtration* of X is a collection of subcomplexes $(K_t)_{t\in\mathbb{R}}$ of X such that $K_s \supseteq K_t$ whenever $s \leq t$. Filtrations (K_t) and (L_t) will be called *equivalent* if there is a constant $\mu \geq 0$ such that $K_{t-\mu} \supseteq L_t \supseteq K_{t+\mu}$ for all t.

The geodesic ray γ defines a filtration of X by the subcomplexes $\{X_{(\gamma,s)} | s \in \mathbb{R}\}$, where $X_{(\gamma,s)}$ is the largest subcomplex of X lying in $h^{-1}(HB_s(\gamma))$. The equivalence class of this filtration only depends on $\gamma(\infty) = e \in \partial M$ since, by Proposition 11.2, $X_{(\gamma',t)} = X_{(\gamma,t+\text{constant})}$ when γ' is asymptotic to γ. And it does not depend on the G-map h since, referring to Proposition 3.1, for any $A \subset M$, $h_2^{-1}(A)$ lies in h_1^{-1}(the t-neighborhood of A), where $t = \sup\{d(h_1(x), h_2(x)) | x \in X^n\}$. Moreover, the property that X be CC^{n-1} over e is, in fact, also independent of the choice of X, so that it is really a property of the action ρ and the point $e \in \partial M$. This is covered by the following more general Invariance Theorem

THEOREM 12.1. *Let G be of type F_n, let Y be a cocompact n-dimensional $(n-1)$-connected rigid G-CW-complex such that the stabilizer of each p-cell is of type F_{n-p}, $p \leq n-1$. Let $h : Y \to M$ be a G-map and let $e \in \partial M$. The property that Y be CC^{n-1} over e is independent of the choices of γ, of Y and of the control function h.*

The proof of Theorem 12.1 carries over, mutatis mutandis, from the proof of Theorem 3.3.

In §10.2 we remarked that the lag λ depends on the horoball $HB_s(\gamma)$. By Proposition 11.2, if γ' is asymptotic to γ then $HB_s(\gamma) = HB_{s'}(\gamma')$ for some s'. We mean that λ is not a function of γ and s but rather depends only on the set $HB_s(\gamma)$.

12.2. CC^{-1} in all directions

Here we prove

THEOREM 12.2. *The action $\rho : G \to \text{Isom}(M)$ is CC^{-1} in all directions $e \in \partial M$ if and only if ρ is cocompact.*

PROOF. We may take $n = 0$ and consider a finite subset $F \subset X^0$ with $GF = X^0$. If ρ is cocompact then there is a compact set $C \subset M$ such that $GC = M$ and $h(F) \cap C \neq \emptyset$. Since C has finite diameter, each horoball must contain a translate of C, hence some point of $h(X^0) = Gh(F)$; i.e. X is (-1)-connected in all directions.

It remains to show that if ρ is not cocompact then there is a horoball $HB_s(\gamma)$ disjoint from $h(X^0)$. There is a sequence (b_m) in M such that $\inf d(h(X^0), b_m) \geq m$ for all m. Let $\epsilon > 0$. There are points $a_m \in h(X^0)$ such that $\inf d(h(X^0), b_m) \leq d(a_m, b_m) \leq \inf d(h(X^0), b_m) + \epsilon$, and translating each b_m with the G-action if necessary we may assume that $a_m \in h(F)$. Pick $a \in h(F)$. Since the metric d is proper the space R_a of all generalized geodesic rays emanating from the base point a is compact. The points $b_m \in M$ are represented in R_a by the unique generalized geodesic rays $\omega_m : [0, \infty) \to M$ starting in a and ending at $\omega_m(\infty) = b_m$. The sequence (ω_m) has a limit point $\gamma \in R_a$, and we may assume it converges to γ. If $\gamma(\infty) \in M$ then the sequence $(d(a, b_m))$ is bounded which contradicts the choice of (b_m). Hence $\gamma(\infty) \in \partial M$, i.e., $\gamma : [0, \infty) \to M$ is a geodesic ray. As (ω_m) converges to γ we find for each $k \in \mathbb{N}$ a number $N(k) \geq k$ with $d(\omega_{N(k)}(k),\ \gamma(k)) < \epsilon$. We abbreviate $c_k := \omega_{N(k)}(k)$ and observe that, as well as $d(\gamma(k), c_k) < \epsilon$, we have also $d(a, c_k) = k$, since c_k is the parameter-k-point on the geodesic segment from a to $b_{N(k)}$ which has length $d(a, b_{N(k)}) + \epsilon \geq d(a_{N(k)}, b_{N(k)}) \geq N(k) \geq k$.

We claim that for each point $c \in h(X^0)$ the inequality

$$d(a, c_k) \leq d(a, a_{N(k)}) + d(c, c_k) + \epsilon$$

holds. Indeed, this follows from

$$\begin{aligned} d(a, b_{N(k)}) &\leq d(a, a_{N(k)}) + d(a_{N(k)}, b_{N(k)}) \\ &\leq d(a, a_{N(k)}) + d(c, b_{N(k)}) + \epsilon \\ &\leq d(a, a_{N(k)}) + d(c, c_k) + d(c_k, b_{N(k)}) + \epsilon \end{aligned}$$

by subtracting $d(c_k, b_{N(k)})$ on either side. It follows that for each $c \in h(X^0)$ and each $k \in \mathbb{N}$

$$\begin{aligned} d(c, \gamma(k)) &\geq d(c, c_k) - d(c_k, \gamma(k)) \\ &\geq d(a, c_k) - d(a, a_{N(k)}) - d(c_k, \gamma(k)) - \epsilon \\ &\geq k - \operatorname{diam} h(F) - 2\epsilon. \end{aligned}$$

This shows that none of the points of $h(X^0)$ is contained in the horoball $HB_{\operatorname{diam} h(F) + 2\epsilon}(\gamma)$. □

CHAPTER 13

Finitary Contractions Towards Endpoints

Throughout this section X is a contractible G-CW-complex with finite stabilizers and cocompact n-skeleton, M is a CAT(0) metric space, an action of G on M by isometries is given, and $h : X \to M$ is a G-equivariant control function. The main technical result in Part I characterizes "CC^{n-1} over $a \in M$" in terms of the existence of a finitary contraction towards a. We now consider finitary contractions towards an endpoint $e \in \partial M$ in connection with the CC^{n-1} property over e.

13.1. Shift and contractions towards $e \in \partial M$

As in §5.2 we consider cellular maps $f : D(f) \to X$, where $D(f)$ is a subcomplex of X. The *shift of f towards $e \in \partial M$* (or *in the direction e*) is defined to be the continuous function $\mathrm{sh}_{f,e} : D(f) \to \mathbb{R}$ given by

$$\mathrm{sh}_{f,e}(x) := \beta_\gamma hf(x) - \beta_\gamma h(x),$$

where γ is a geodesic ray representing e. By Proposition 11.2(a) this is independent of the choice of γ. By the definition of β_γ and the triangle inequality the shift is bounded by the displacement function

$$|\mathrm{sh}_{f,e}(x)| \le \alpha_f(x), \tag{13.1}$$

hence, just as for the shift towards $a \in M$, the shift function has a global bound $\|f\|$ if f is a bounded map[1].

By the *guaranteed shift towards $e \in \partial M$* we mean

$$\mathrm{gsh}_e(f) := \inf \mathrm{sh}_{f,e}(D(f)).$$

Just as in §5.4 we find

$$\mathrm{gsh}_{ge}(gf) = \mathrm{gsh}_e(f), \qquad \text{all } g \in G.$$

A cellular map $\phi : X \to X$ is said to be a *contraction towards $e \in \partial M$* (or *in the direction e*) if $\mathrm{gsh}_e(\phi) > 0$.

13.2. From contractions to CC^{n-1}

The endpoint versions of the propositions in §5.3 have easier statements and proofs. Corresponding to Proposition 5.3 we have the observation that

$$\mathrm{gsh}_e(\phi^m) \ge m\,\mathrm{gsh}_e(\phi) \tag{13.2}$$

for each contraction $\phi : X \to X$ and each $m \in \mathbb{N}$. Corresponding to Proposition 5.5 we have

[1] Recall from §5.1 that $\alpha_f(x) := d(h(x), hf(x))$ and $||f|| := \sup \alpha_f(D(f))$.

PROPOSITION 13.1. *If X^n admits a finitary contraction $\phi : X^n \to X^n$ towards $e \in \partial M$ and $\epsilon := \mathrm{gsh}_e(\phi)$ then there is a number $\lambda \geq 0$ and a cellular deformation $\psi : X^n \times [0, \infty) \to X^{n+1}$ satisfying the Lipschitz condition*

$$\beta_\gamma h\psi(x, s_2) - \beta_\gamma h\psi(x, s_1) \geq (s_2 - s_1)\varepsilon - \lambda$$

whenever $s_1 \leq s_2$, $x \in X^n$, and γ is a geodesic ray with $\gamma(\infty) = e$.

PROOF. As in the proof of Proposition 5.5 we define $\psi_0 : X^n \times I \to X^{n+1}$ to be a finitary homotopy $\mathrm{Id}_{X^n} \simeq \phi$ and take $\psi(x,t) := \psi_0(\phi^m(x), s)$, where $x \in X^n$, $t \in [0, \infty)$, and m is an integer with $s = t - m \in I$. Define $\mathrm{sh}_{\psi_{0,e}}(x) := \sup_{t \in I} \mathrm{sh}_{\psi_{0,e}(\cdot,t)}(x)$. This time we note that for $y := (\phi^m(x), t_1)$ and $z := (\phi^m(x), t_2)$ with $0 \leq t_1 \leq t_2 \leq 1$ and $x \in X^n$,

$$\begin{aligned} |\beta_\gamma h\psi_0(y) - \beta_\gamma h\psi_0(z)| &\leq |\beta_\gamma h\psi_0(y) - \beta_\gamma h(\phi^m(x))| + |\beta_\gamma h(\phi^m(x)) - \beta_\gamma h\psi_0(z)| \\ &\leq |\mathrm{sh}_{\psi_0,e}(\phi^m(x))| + |\mathrm{sh}_{\psi_0,e}(\phi^m(x))| \\ &\leq 2\|\psi_0\|. \end{aligned}$$

If $s_i := m_i + t_i$ as above, with $s_1 \leq s_2$, it follows that

$$\begin{aligned} \beta_\gamma h\psi(x, s_2) - \beta_\gamma h\psi(x, s_1) &= \beta_\gamma h\psi_0(\phi^{m_2}(x), t_2) - \beta_\gamma h\psi_0(\phi^{m_1}(x), t_1) \\ &\geq \beta_\gamma h\psi_0(\phi^{m_2}(x), 1) - \beta_\gamma h\psi_0(\phi^{m_1}(x), 0) - 4\|\psi_0\| \\ &= \beta_\gamma h\phi^{m_2+1}(x) - \beta_\gamma h\phi^{m_1}(x) - 4\|\psi_0\| \\ &= \mathrm{sh}_{\phi^{m_2-m_1+1},e}(\phi^{m_1}(x)) - 4\|\psi_0\| \\ &\geq \mathrm{gsh}_e(\phi^{m_2-m_1+1}) - 4\|\psi_0\| \\ &\geq (m_2 - m_1 + 1)\varepsilon - 4\|\psi_0\|, \quad \text{by (13.2)} \\ &\geq (s_2 - s_1)\varepsilon - 4\|\psi_0\|. \end{aligned}$$

So we can choose λ to be $4\|\psi_0\|$. □

By a straightforward adaptation of the proof of Theorem 5.6 we get

THEOREM 13.2. *If X is contractible with finite stabilizers and cocompact n-skeleton then the existence of a finitary contraction $\phi : X^n \to X^n$ towards $e \in \partial M$ implies that X is CC^{n-1} over e with constant lag $\lambda = 4\|\psi_0\|$, where ψ_0 is any finitary homotopy $\mathrm{Id}_{X^n} \simeq \phi$.* □

There is a subtle variation of Theorem 13.2 which did not arise in the parallel case in Chapter 5 but will be needed in the inductive proof of the main result (Proposition 14.3).

COROLLARY 13.3. *The existence of a finitary contraction $\phi : X^n \to X^n$ towards $e \in \partial M$ also implies that if X has cocompact $(n+1)$-skeleton and is CC^n over e then it is so with constant lag $\lambda = 4\|\psi_0\|$.*

PROOF. Let $\gamma(\infty) = e$ as before. Assume X is CC^n over e with lag $\lambda_0(t)$ and consider a map $f : S^n \to X_{(e,t)}$. Put $\varepsilon := \mathrm{gsh}_e(\phi)$. For any $r \geq 0$ we can compose the deformation ψ of Proposition 13.1 with the map $(f \times \mathrm{Id}) : S^n \times [0, r] \to X_{(\gamma,t)} \times [0, r]$. By the Lipschitz condition of ψ this yields a homotopy $H : S^n \times [0, r] \to X_{(\gamma,t-\lambda)}$ between f and a map $f' = H(\cdot, r) : S^n \to X_{(\gamma,t+r\epsilon-\lambda)}$. By the CC^n-property of X over e, with lag $\lambda_0(t)$, we can choose r so large that $t + r\epsilon - \lambda - \lambda_0(t + r\epsilon - \lambda) > t - \lambda$; this is because $s - \lambda_0(s) \to \infty$ as $s \to \infty$. Then f' extends to a map $\tilde{f}' : B^{n+1} \to X_{(\gamma,t-\lambda)}$. Now, the annulus $S^n \times [0, r]$ and the ball B^{n+1} can be glued together along $S^n \times \{r\}$ to make a topological $(n+1)$-ball $\tilde{B}$, and the union $\tilde{f} = H \cup \tilde{f}'$ is a map $\tilde{f} : \tilde{B} \to X_{(\gamma,t-\lambda)}$ extending f. □

13.3. Passing to the closure of G-orbits

It is clear that if $\phi : X^n \to X^n$ is a finitary contraction towards $e \in \partial M$ then $g\phi$ is a finitary contraction towards ge, so that, by Theorem 13.2, X is CC^{n-1} over each e' in the G-orbit of e, with a uniform constant lag λ. We improve this by establishing

PROPOSITION 13.4. *The existence of a finitary contraction $\phi : X^n \to X^n$ towards $e \in \partial M$ implies that X is CC^{n-1} over every e' in $\mathrm{cl}_{\partial M}(Ge)$, the closure of the G-orbit of e. Moreover, if ψ_0 is a finitary homotopy $\mathrm{Id}_{X^n} \simeq \phi$ then any number $\lambda' > 4\|\psi_0\|$ is a constant lag, uniform for all $e' \in \mathrm{cl}_{\partial M}(Ge)$.*

In the proof of Proposition 13.4 we will need the following

LEMMA 13.5. *Let M be a $\mathrm{CAT}(0)$ space and ε, r positive numbers. Then any number $R > r(1 + 2r/\varepsilon)$ has the property that whenever $\gamma, \gamma' : [0, \infty) \to M$ are geodesic rays starting at $c \in M$ then*

$$|\beta_\gamma(p) - \beta_{\gamma'}(p)| < 2\varepsilon + d(\gamma(R), \gamma'(R))$$

for every $p \in B_r(c)$.

PROOF. Applying Lemma 6.4 with $a = \gamma(t), b = \gamma(R)$ and $p' = c$ yields

$$|d(c, \gamma(t)) - d(p, \gamma(t)) - d(c, \gamma(R)) + d(p, \gamma(R))| < \varepsilon.$$

Hence passing to the limit $t \to \infty$ we find

$$|\beta_\gamma(p) - R + d(p, \gamma(R))| < \varepsilon,$$

and similarly for γ'. Subtracting the two inequalities and using the triangle inequality yields the lemma. □

PROOF (OF PROPOSITION 13.4). We use the notation of the proof of Proposition 13.1. The proof of Theorem 13.2 starts with $0 \le p \le n-1$, a map $f : S^p \to X^p$, its extension to $f_1 : B^{p+1} \to X^{p+1}$, and a number t such that $f(S^p) \subset X_{(\gamma,t)}$. The deformation ψ is used to move f_1 to $f_1' : B^{p+1} \to X^{p+1}$, where $f_1' := \psi(f_1(\cdot), T)$, T being sufficiently large that $f_1'(B^{p+1}) \subseteq X_{(\gamma,t-\lambda)}$. The Lipschitz condition on ψ shows that if $\epsilon := \mathrm{gsh}_e(\phi)$ then $T = \mathrm{diam}\; hf_1(B^{p+1})/\epsilon$ will do. Let r be the diameter over M of the $[0,T]$ track of $f_1(B^{p+1})$, i.e. $r := \mathrm{diam}\; h\psi(f_1(B^{p+1}) \times [0,T])$. The final map $\tilde{f} : B^{p+1} \to X^p$ satisfies $\mathrm{diam}\; h\tilde{f}(B^{p+1}) \le r$ and $\tilde{f}(B^{p+1}) \subset X_{(\gamma,t-\lambda)}$.

The numbers $\lambda = 4\|\psi_0\|$, $\epsilon = \mathrm{gsh}_e(\phi)$, T and r remain unchanged when e and ϕ are replaced by G-translates ge and $g\phi$. This shows that for every $g \in G$ the given map $f : S^p \to X^p$ can be extended to a map $\tilde{f}_g : B^{p+1} \to X^{p+1}$ such that (i) $\mathrm{diam}\; h\tilde{f}_g(B^{p+1}) \le r$ and (ii) $f(S^p) \subset X_{(g\gamma,s)}$ implies $\tilde{f}_g(B^{p+1}) \subset X_{(g\gamma,s-\lambda)}$. In words, (ii) says that the extension $\tilde{f}_g$ of f has lag λ with respect to ge.

Now choose a base point $c \in hf(S^p)$. For all $g \in G$ we have $h\tilde{f}_g(B^{p+1}) \subset B_r(c)$. Let $e' \in \mathrm{cl}(Ge)$ be represented by a geodesic ray $\gamma' : [0, \infty) \to M$ with $\gamma'(0) = c$. Let $\nu > 0$ be an arbitrary small number, choose $R > r(1+2r/\nu)$, and choose $g \in G$ such that if ge is represented by the geodesic ray $\overline{\gamma} : [0, \infty) \to M$ with $\overline{\gamma}(0) = c$, then $d(\gamma'(R), \overline{\gamma}(R)) < \nu$. Lemma 13.5 shows that

$$|\beta_{\gamma'}(u) - \beta_{\overline{\gamma}}(u)| < 3\nu, \text{ for all } u \in B_r(c).$$

This shows that the extension $\tilde{f}_g : B^{p+1} \to X^{p+1}$ of $f : S^p \to X^p$ has lag $\lambda + 3\nu$ with respect to e'. □

13.4. Contractions in a set of directions $E \subseteq \partial M$

We conclude this section with a further improvement on Proposition 13.4 and Corollary 13.3.

PROPOSITION 13.6. *Let $\mathcal{F} : X^n \to X^n$ be a locally finite G-sheaf and let $E \subseteq \partial M$ be a closed set of endpoints with the property that for each $e \in E$ the sheaf $\mathcal{F}$ has a cross section $\phi_e : X^n \to X^n$ with $\inf\{gsh_e(\phi_e) | e \in E\} = \varepsilon > 0$. Then there is a number λ such that the following hold:*

(a) *X is CC^{n-1} over each $e' \in \mathrm{cl}_{\partial M}(GE)$ with constant lag λ, uniform for all $e' \in \mathrm{cl}_{\partial M}(GE)$.*

(b) *If X has cocompact $(n+1)$-skeleton and is CC^n over $e' \in GE$ then it is so with lag λ.*

PROOF. The lag for an individual endpoint $e \in E$ which we found in Theorem 13.2 and Corollary 13.3 was $4||\psi_{0,e}||$ where $\psi_{0,e}$ was a finitary homotopy $\mathrm{Id}_{X^n} \simeq \phi_e$. Such a homotopy is obtained as a cross section of a homotopy of sheaves $\mathcal{H} : \mathcal{F}_0 \simeq \mathcal{F}$, where $\mathcal{F}_0$ stands for the sheaf consisting of all identity maps Id_D with $D \subseteq X^n$ the domain of a member of $\mathcal{F}$; see Propositions 4.8 and 4.10. Since $\psi_{0,e}$ is a cross section of $\mathcal{H}$, we have $||\psi_{0,e}|| \leq ||\mathcal{H}|| := \sup\{||H|| \mid H \in \mathcal{H}\}$. Therefore $4||\mathcal{H}||$ is a uniform lag for all $e \in E$, and Proposition 13.4 shows that any number $\lambda > 4||\mathcal{H}||$ is a uniform lag for all $e' \in \cup\{\mathrm{cl}\ Ge \mid e \in E\}$. Since E is closed in ∂M this set is $\mathrm{cl}_{\partial M} GE$. □

CHAPTER 14

From CC^{n-1} over Endpoints to Contractions

Let X, M and h be as in Chapter 13. We aim to construct finitary contractions from CC^{n-1} assumptions over endpoints.

14.1. Vertex shift and defect of sheaves

Let $\mathcal{F} : X \rightsquigarrow X$ be a locally finite homotopically closed G-sheaf. In §5.4 and §5.5 we introduced the maximal guaranteed vertex shift $\mu_a(\mathcal{F}|\sigma)$ and the defect $d_a(\mathcal{F}|\sigma)$ on a cell σ towards a point $a \in M$. The same definitions apply with $a \in M$ replaced by an endpoint $e \in \partial M$.

Guaranteed vertex shift and defect in the direction $e \in \partial M$ are used to control the shift of a cross section $\phi : X \to X$ of $\mathcal{F}$ on σ in the direction e. Just as in §5.5 one introduces the total defect of $\mathcal{F}$ on σ in the direction e, $\delta_e(\mathcal{F}|\sigma)$, and one proves (compare Proposition 5.7 and Remark 5.8):

PROPOSITION 14.1. *Let $e \in \partial M$. Then every locally finite homotopically closed G-sheaf $\mathcal{F} : X \rightsquigarrow X$ has a cross section $\phi : X \to X$ with*

$$\mathrm{sh}_{\phi,e}(x) \geq \mu_e(\mathcal{F}|\sigma) - \delta_e(\mathcal{F}|\sigma)$$

for each cell σ of X and each $x \in \sigma$; and ϕ can be chosen to be a G_e-map. □

14.2. Controlled embedding of sheaves into homotopically closed sheaves

We shall need the following embedding result

PROPOSITION 14.2. *Let E be a closed and G-invariant subset of ∂M. Assume X is CC^{n-1} in all directions $e \in E$ with uniform constant lag $\lambda \geq 0$ depending only on E. Then for every $\lambda' > \lambda$, any locally finite G-sheaf $\mathcal{F} : X^n \rightsquigarrow X^n$ can be embedded in a homotopically closed locally finite G-sheaf $\tilde{\mathcal{F}}$ such that $d_e(\tilde{\mathcal{F}}|\sigma) \leq \lambda' + \mathrm{diam}\ hC(\sigma)$, for all cells σ of X^n, and all $e \in E$.*

PROOF. The argument is parallel to those in §6.2 and §6.4, in fact considerably simpler since the technicalities of §6.3 do not arise. Just as in §6.2 one starts by fixing an n-cell σ of X and a cellular map $f : C(\dot{\sigma}) \to X$ and one proves that, given any $\varepsilon > 0$, a finite set $S(\sigma)$ of cellular maps $C(\sigma) \to X$ extending f can be chosen in such a way that for each geodesic ray γ with $\gamma(\infty) = e \in E$ there exist $\tilde{f}_e \in S(\sigma)$ with

$$\mathrm{gsh}_\gamma(\tilde{f}_e) > \mathrm{gsh}_\gamma(f) - \mathrm{diam}\ hC(\sigma) - \lambda - \varepsilon.$$

Note that we have used the assumption that E is compact and the fact that the Busemann function β_γ is continuous in γ (Proposition 11.2).

We can now prove Proposition 14.2 by induction on n. When $n = 0$ the condition on the defects $d_e(\mathcal{F}|\sigma)$ is empty so we can refer to Proposition 4.5. By induction, assume that $\mathcal{F}^{n-1} := \mathcal{F} \mid X^{n-1} : X^{n-1} \rightsquigarrow X^{n-1}$ is homotopically closed with $d_e(\mathcal{F}^{n-1}|\tau) \leq \lambda' + \text{diam } hC(\tau)$ for every cell τ of X^{n-1} and all $e \in E$. In order to embed $\mathcal{F}$ in a homotopically closed sheaf we adjoin the sets $S(\sigma)$ constructed above to $\mathcal{F}$. The set $S(\sigma)$ depends on σ; but for the G-translates $g\sigma$ of σ we do not have to choose $S(g\sigma)$ anew but can put $S(g\sigma) := gS(\sigma)$. This is because $\text{gsh}_{ge}(gf) = \text{gsh}_e(f)$; see §13.1. In this way we have embedded $\mathcal{F}$ in a sheaf $\tilde{\mathcal{F}} : X^n \rightsquigarrow X^n$ which is homotopically closed, locally finite and a G-sheaf; and $\tilde{\mathcal{F}}$ does satisfy the required condition on the defect. □

14.3. Obtaining contractions towards end points

Here is the converse to Proposition 13.6.

PROPOSITION 14.3. *If $E \subseteq \partial M$ is a closed G-invariant subset and X is CC^{n-1} over every $e \in E$ then there is a locally finite homotopically closed G-sheaf $\mathcal{F} : X^n \rightsquigarrow X^n$ which admits, for each $e \in E$, a cross section $\phi_e : X^n \to X^n$ with* $\inf\{\text{gsh}_e(\phi_e)|e \in E\} > 0$.

PROOF. First we consider the case $n = 0$. For each vertex v of a finite fundamental domain $F \subseteq X^n$ and each end point $e \in E$ we use the CC^{-1} hypothesis to pick a map $f_e^v : \{v\} \to X^0$ with $\text{gsh}_e(f_e^v) = hf_e^v(v) - h(v) > 0$. For $e' \in E$ sufficiently close to e, $\text{gsh}_{e'}(f_e^v) > 0$. Since E is compact we find a finite sheaf $\mathcal{F}_0 : F^0 \rightsquigarrow X^0$ with $\mu_e(\mathcal{F}_0|v) > 0$ for all $e \in E$ and all $v \in F^0$. Since F^0 and the stabilizers of the vertices $v \in F^0$ are finite there are only finitely many group elements $g \in G$ with $gF^0 \cap F^0 \neq \emptyset$. By adding to $\mathcal{F}_0$ the maps gf for all $f \in \mathcal{F}_0$ and all $g \in G$ with $gD(f) \in F^0$ we find a finite G-saturated sheaf $\mathcal{F}_0' : F \rightsquigarrow X^0$ (in the sense of §4.2).

By Propositions 4.1 and 4.3 $\mathcal{F}^0 := G\mathcal{F}_0' : X^0 \rightsquigarrow X^0$ is a locally finite G-sheaf. Since it contains $\mathcal{F}_0'$ and since $\mu_{ge}(\mathcal{F}^0|gv) = \mu_e(\mathcal{F}^0|v)$ we have $\mu_e(\mathcal{F}^0|x) > 0$ for all $e \in E$ and all $x \in X^0$ – in fact $\inf \mu_e(\mathcal{F}^0|X^0) > 0$. The required cross section ϕ_e of $\mathcal{F}^0$ exists by Proposition 14.1.

We proceed by induction on n, assuming that a homotopically closed locally finite G-sheaf $\mathcal{F} : X^{n-1} \rightsquigarrow X^{n-1}$ has already been constructed, with cross sections $\phi_e : X^{n-1} \to X^{n-1}$ contracting towards $e \in E$, and $\varepsilon := \inf\{\text{gsh}_e(\phi_e)|e \in E\} > 0$. The assumptions of Proposition 13.6(b) are then fulfilled (with n replaced by $n-1$), so there is a uniform constant lag for the CC^{n-1}-assumptions on X in all directions $e \in E$. In this situation Proposition 14.2 applies and yields a number D with the property that every locally finite G-sheaf $\mathcal{F}' : X^n \rightsquigarrow X^n$ can be embedded into a locally finite homotopically closed G-sheaf $\tilde{\mathcal{F}}' : X^n \rightsquigarrow X^n$ with $d_e(\tilde{\mathcal{F}}'|\sigma) \leq D$ for all $e \in E$, and all cells σ of X^n.

We may alter the domain of $\mathcal{F}$ and write $\mathcal{F} : X^n \rightsquigarrow X^n$, noting that $\mathcal{F} \mid X^{n-1} : X^{n-1} \rightsquigarrow X^{n-1}$ (but perhaps not $\mathcal{F}$) is homotopically closed. Let m be a positive integer such that $m \cdot \varepsilon > D$. The m-th iterate $\mathcal{F}^{(m)} : X^n \rightsquigarrow X^n$ is a locally finite G-sheaf and thus, by the previous paragraph, can be embedded into a locally finite homotopically closed G-sheaf $\tilde{\mathcal{F}} : X^n \rightsquigarrow X^n$ with $d_e(\tilde{\mathcal{F}}|\sigma) \leq D$ for all $e \in E$ and all cells σ of X^n. Then the maps $\phi_e^m : X^{n-1} \to X^{n-1}$ are cross sections of $\tilde{\mathcal{F}}$ and, by (13.2), $\text{gsh}_e \phi_e^m \geq m\text{gsh}_e(\phi_e) \geq m \cdot \varepsilon$. From the definition of the defect $d_e(\tilde{\mathcal{F}}|\sigma)$ it is clear that each of the cross sections $\phi_e^m : X^{n-1} \to X^{n-1}$ can be extended to a cross section $\tilde{\phi}_e : X^n \to X^n$ of the homotopically closed sheaf $\tilde{\mathcal{F}}$ with

$$\begin{aligned}\mathrm{gsh}_e(\tilde{\phi}_e|C(\sigma)) &\geq \mathrm{gsh}_e(\phi_e^m|C(\sigma)) - d_e(\tilde{\mathcal{F}}|\sigma)\\ &\geq m\varepsilon - D > 0,\end{aligned}$$

for all cells σ of X^n. □

14.4. The main technical results

THEOREM 14.4. *If X is a contractible G-CW-complex with finite cell stabilizers and cocompact n-skeleton and if G acts on M by isometries then the following are equivalent for $e \in \partial M$.*

(i) *X^n admits a G-finitary contraction towards e*
(ii) *X is CC^{n-1} in all directions $e' \in \mathrm{cl}_{\partial M}(Ge)$*
(iii) *There is a locally finite homotopically closed G-sheaf $\mathcal{F} : X^n \rightsquigarrow X^n$ and a number $\varepsilon > 0$ with*

$$\mu_e(\mathcal{F}|\sigma) - \delta_e(\mathcal{F}|\sigma) \geq \varepsilon$$

for all cells σ of X^n.

ADDENDUM. *If* (i)–(iii) *hold then there is a constant lag $\lambda \geq 0$ uniform for all $e' \in \mathrm{cl}(Ge)$ in* (ii).

PROOF. (i) $\Rightarrow$ (ii) is covered by Proposition 13.4 which yields a uniform constant lag λ as required in the Addendum. (ii) $\Rightarrow$ (i) is covered by Proposition 14.3. The implication (iii) $\Rightarrow$ (i) is clear from Proposition 14.1. It remains to show that (i) implies (iii). We know that (i) implies (ii) with uniform constant lag λ as described in the Addendum. Let $\lambda' > \lambda$ and let $L > \lambda' + \mathrm{diam}\ hC(\sigma)$ for all cells σ of X^n. Let $\phi : X^n \to X^n$ be a finitary contraction towards e. Then $\mathrm{gsh}_e(\phi) > 0$, and for some $m \geq 1$ ϕ^m is a finitary contraction towards e with $\mathrm{gsh}_e(\phi^m) > L$; this uses Proposition 4.9. Then $\mathcal{F}' := G\ \mathrm{Res}(\phi^m)$ is a locally finite G-sheaf with $\mu_e(\mathcal{F}' \mid \sigma) > L$ for all cells σ of X^n. By Proposition 14.2, using (ii) with uniform constant lag λ on $\mathrm{cl}_{\partial M}(Ge)$, $\mathcal{F}'$ can be embedded in a homotopically closed locally finite G-sheaf $\mathcal{F}$ with $d_e(\mathcal{F} \mid \sigma) \leq \lambda' + \mathrm{diam}\ hC(\sigma) < L$ for all cells σ of X^n. Since $\mu_e(\mathcal{F} \mid \sigma) > L$ for all σ, (iii) holds. □

REMARK. For any cell σ of X and any $\mathcal{F}$ as above, the numbers $\mu_e(\mathcal{F} \mid \sigma)$ and $\delta_e(\mathcal{F} \mid \sigma)$ depend continuously on $e \in \partial M$; this follows from Proposition 11.2; compare §5.4. Hence the inequality $\mu_e(\mathcal{F} \mid \sigma) - \delta_e(\mathcal{F} \mid \sigma) > \frac{\epsilon}{2}$ which comes from (iii) of Theorem 14.4 for given e implies $\mu_{e'}(\mathcal{F} \mid \sigma) - \delta_{e'}(\mathcal{F} \mid \sigma) > \frac{\epsilon}{2}$ for all e in some neighborhood of e. However, since there are infinitely many σ to be considered one cannot conclude that $\overset{\circ}{\Sigma}{}^n(\rho)$ is open in ∂M with respect to the cone topology. Counterexamples are found in §§10.4, 10.7(B) and 10.7(C).

THEOREM 14.5. *If X is a contractible G-CW-complex with finite cell stabilizers and cocompact n-skeleton and if G acts on M by isometries then the following are equivalent for a G-invariant closed subset $E \subseteq \partial M$.*

(i) *X^n admits G-finitary contractions $\phi_e : X^n \to X^n$ towards each $e \in E$.*
(ii) *X is CC^{n-1} in all directions $e \in E$.*
(iii) *There is a locally finite homotopically closed G-sheaf $\mathcal{F} : X^n \rightsquigarrow X^n$ and a number $\varepsilon > 0$ with*

$$\mu_e(\mathcal{F}|\sigma) - \delta_e(\mathcal{F}|\sigma) \geq \varepsilon$$

for all cells σ *of* X^n *and all* $e \in E$.

ADDENDUM. *If* (i) – (iii) *hold then more is true. There is a locally finite homotopically closed* G*-sheaf* $\mathcal{F} : X^n \rightsquigarrow X^n$ *such that the maps* ϕ_e *in* (i) *can be chosen to satisfy* $\mathrm{Res}(\phi_e) \subset \mathcal{F}$ *for all* e *and* $\inf\{\mathrm{gsh}_e(\phi_e) | e \in E\} > 0$*; and* (ii) *holds with a uniform constant lag* λ *in all directions* $e \in E$.

PROOF. (i) $\Leftrightarrow$ (ii) follows from Theorem 14.4. (iii) $\Rightarrow$ (i) is immediate from Proposition 14.1. It remains to prove that (i) and (ii) imply (iii). Assuming (ii), Proposition 14.3 gives the stronger form of (i) in the Addendum; and this implies the stronger form of (ii) in the Addendum (uniform constant lag λ) by Proposition 13.6. Thus in proving (i) $\Rightarrow$ (iii) we may assume the stronger forms of (i) and (ii). The proof runs parallel to the (i) $\Rightarrow$ (iii) part of Proposition 14.4. Let $L > \lambda' + \mathrm{diam}\, hC(\sigma)$ for all cells σ of X^n. For each $e \in E$ let $\phi_e : X^n \to X^n$ be a finitary contraction towards e, each ϕ_e being a cross section of the sheaf $\mathcal{F}$ in the Addendum, and $\inf\{\mathrm{gsh}_e(\phi_e) \mid e \in E\} > 0$. By Propositions 4.7 and 4.9, each iterate ϕ_e^m is a finitary contraction towards e and is a cross section of the m-fold composite sheaf $\mathcal{F}' := \mathcal{F}^{(m)}$. This is a locally finite G-sheaf and, if m is large enough, $\mu_e(\mathcal{F}' \mid \sigma) > L$ for all cells σ of X^n and all $e \in E$. The rest of the proof is similar to the corresponding proof in Theorem 14.4. □

CHAPTER 15

Proofs of Theorems E-H

15.1. Dynamical characterization of $\overset{\circ}{\Sigma}{}^n(\rho)$

PROOF OF THEOREM E. This is the equivalence of (i) and (ii) in Theorem 14.4. $\square$

15.2. Openness using the Tits distance topology

Recall from §5.1 that the norm of a cellular map $\phi : X^n \to X^n$ is $||\phi|| := \sup\{d(h(x), h\phi(x)) \mid x \in X^n\}$, and that ϕ is said to be bounded if $||\phi|| < \infty$. Recall from §10.3 that ϕ is a contraction towards $e \in \partial M$ if there is a number $\epsilon > 0$ such that $\mathrm{sh}_{\phi,e}(x) \geq \epsilon$ for all $x \in X^n$ (see §13.1 for $\mathrm{sh}_{\phi,e}(x)$).

PROPOSITION 15.1. *Let ϕ be a bounded contraction towards $e \in \partial M$. There is a neighborhood N of e in $(\partial M, Td)$ such that ϕ is also a bounded contraction towards every $e' \in N$.*

PROOF. Let γ and γ' be geodesic rays representing e and $e' \in \partial M$ such that $\gamma(0) = \gamma'(0)$. Let $\alpha(t)$ be the angle at $\gamma(0)$ in the geodesic triangle whose other vertices are $\gamma(t)$ and $\gamma'(t)$. With notation as in §11.4, it is shown in [**BrHa**, III 3.4] that $\alpha(t) \leq \angle(e, e')$ and $\lim_{t\to\infty} \alpha(t) = \angle(e, e')$. By Euclidean geometry $\sin \frac{\alpha(t)}{2} = \frac{d(\gamma(t),\gamma'(t))}{2t}$, so

$$d(\gamma(t), \gamma'(t)) \leq 2t \sin\left(\frac{\angle(e, e')}{2}\right).$$

The right hand side of this inequality is independent of the base point $\gamma(0)$.

Let $\epsilon > 0$ be such that $\mathrm{sh}_{\phi,e}(x) \geq \epsilon$ for all $x \in X^n$. We apply Lemma 13.5 (with $\frac{\epsilon}{6}$ replacing ϵ in that lemma, and $r = ||\phi||$). Letting $R = ||\phi||(1 + \frac{12||\phi||}{\epsilon}) + \epsilon$, that lemma implies that for any $x \in X^n$ and any $p \in B_{||\phi||}(h(x))$ we have:

$$|\beta_\gamma(p) - \beta_{\gamma'}(p)| < \frac{\epsilon}{3} + d(\gamma(R), \gamma'(R))$$

where $\gamma(0) = \gamma'(0) = h(x)$. Thus if e' is chosen so that $\angle(e, e') < 2\arcsin(\frac{\epsilon}{6R})$ we find

$$\begin{aligned}|\mathrm{sh}_{\phi,e}(x) - \mathrm{sh}_{\phi,e'}(x)| &= |\beta_\gamma h\phi(x) - \beta_{\gamma'} h\phi(x)| \\ &\leq \frac{\epsilon}{3} + \frac{\epsilon}{3R}R = \frac{2\epsilon}{3}.\end{aligned}$$

This shows that $\mathrm{sh}_{\phi,e'}(x) > \frac{\epsilon}{3}$ for all $x \in X^n$. $\square$

Parallel to Proposition 7.3 we have (when h_ρ is chosen continuously in ρ)

PROPOSITION 15.2. *Let $\mathcal{F} : X \rightsquigarrow X$ be a locally finite homotopically closed G-sheaf. Then $\mu_e^\rho(\mathcal{F}|\sigma)$ and $\delta_e^\rho(\mathcal{F}|\sigma)$ are jointly continuous in the variables $(\rho, e) \in \mathrm{Hom}(G, \mathrm{Isom}(M)) \times \partial M$, where ∂M has the cone topology.*

Of course this remains true if ∂M has the finer Tits distance topology.

PROOF OF THEOREM F. For fixed ρ, openness in e follows from Proposition 15.1 and Theorem 14.4. If ρ is permitted to vary and h_ρ is chosen continuously in ρ, the same proof works, in view of Proposition 15.2. □

15.3. Openness in ρ using the cone topology on ∂M

PROOF OF THEOREM G. Let $\rho \in \mathrm{Hom}(G, \mathrm{Isom}(M, E))$ be such that one and hence all of the Conditions (i)-(iii) of Theorem 14.7 hold. Since $\mu_{ge}(\mathcal{F} \mid g\sigma) = \mu_e(\mathcal{F} \mid \sigma)$ and $\delta_{ge}(\mathcal{F} \mid g\sigma) = \delta_e(\mathcal{F} \mid \sigma)$ for all $g \in G$ the inequality in Condition (iii) need only be considered for a (finite) set of representatives of the G-orbits of cells in X^n. The parameter range $e \in E$ is compact and hence the condition (with ϵ replaced by $\frac{\epsilon}{2}$) remains true if ρ is subject to a small perturbation, in view of Proposition 15.2. □

15.4. Endpoints versus points in M

Theorem H follows from the Invariance Theorem 12.1 and

THEOREM 15.3. *Let X be a contractible G-CW-complex with finite cell stabilizers and cocompact n-skeleton. If $h : X \to M$ is an equivariant control function over an almost geodesically complete* CAT(0)*-space M then the following are equivalent*
- (i) *X is CC^{n-1} in all directions $e \in \partial M$*
- (ii) *X is CC^{n-1} over M.*

PROOF. (ii) $\Rightarrow$ (i) is easy in view of Theorem 6.8; it is left to the reader. We prove (i) $\Rightarrow$ (ii). Let $\mathcal{F} : X^n \rightsquigarrow X^n$ be the sheaf given by (iii) of Theorem 14.5 for $E = \partial M$. Let $r := ||\mathcal{F}||$, let ν be an arbitrarily small positive number and choose $S > r(1 + 2r/\nu)$. Let $\mu \geq 0$ be the number given by the definition of "almost geodesically complete" in §10.5. We may assume $\nu \leq 1$ and $\mu \geq 1$ and put $R := \mu S/\nu$. Now we fix a base point $a \in M$ and consider any $x \in X^n$ with $d(a, h(x)) \geq R$. By assumption there is a geodesic ray $\gamma : [0, \infty) \to M$ with $\gamma(0) = h(x)$ and $d(a, \gamma(T)) \leq \mu$ for some $T \in [0, \infty)$. We put $e := \gamma(\infty)$ and we compare the shifts $\mathrm{sh}_{f,a}(x)$ and $\mathrm{sh}_{f,e}(x)$ for all $f \in \mathcal{F}$.

Since $R \geq S > r(1+2r/\nu)$ Lemma 6.4 applies. We apply it, in fact, twice, each time taking $c = p = h(x)$ and $p' = hf(x)$. The first application is for the point $\gamma(t) \notin B_S(c)$ and yields

$$|\mathrm{sh}_{f,\gamma(t)}(x) - \mathrm{sh}_{f,\gamma(S)}(x)| < \nu.$$

Taking the limit $t \to \infty$ this becomes

$$|\mathrm{sh}_{f,e}(x) - \mathrm{sh}_{f,\gamma(S)}(x)| < \nu. \tag{15.1}$$

The second application is for the point $a \notin B_S(c)$ and yields

$$|\mathrm{sh}_{f,a}(x) - \mathrm{sh}_{f,b}(x)| < \nu, \tag{15.2}$$

where b is the point on the geodesic segment from c to a with $d(c, b) = S$. In order to get a bound on $d(\gamma(S), b)$ we have to compare the two isosceles triangles $(a, c, \gamma(d(a, c)))$ and $(b, c, \gamma(S))$ with the corresponding Euclidean ones. Since $d(a, \gamma(d(a, c))) \leq d(a, \gamma(T)) + |T - d(a, c)| \leq 2d(\alpha, \gamma(T)) \leq 2\mu$ we find $d(\gamma(S), b)/S \leq 2\mu/d(a, c) \leq 2\mu/R$, and hence $d(\gamma(S), b) \leq 2\nu$. By (5.7) it follows that

$$|\mathrm{sh}_{f,\gamma(S)}(x) - \mathrm{sh}_{f,b}(x)| \leq 4\nu, \tag{15.3}$$

and hence, from the conjunction of (15.1) - (15.3),

$$|\mathrm{sh}_{f,e}(x) - \mathrm{sh}_{f,a}(x)| \leq 6\nu. \tag{15.4}$$

We use (15.4) to compare defect and maximal guaranteed vertex shift towards $a \in M$ and $e \in \partial M$. Choosing ν sufficiently small compared with ε (as given in (iii) of Theorem 14.5) we find – details left to the reader –

$$|\mu_e(\mathcal{F} \mid \sigma) - \mu_a(\mathcal{F} \mid \sigma)| < \frac{\varepsilon}{3}$$

and

$$|\delta_e(\mathcal{F} \mid \sigma) - \delta_a(\mathcal{F} \mid \sigma)| < \frac{\varepsilon}{3}$$

and therefore

$$|\mu_a(\mathcal{F} \mid \sigma) - \delta_a(\mathcal{F} \mid \sigma)| \geq \frac{\varepsilon}{3}$$

for all cells σ of X^n with $h(C(\sigma)) \cap B_R(a) = \emptyset$. By Theorem 6.8 this establishes that X is CC^{n-1} over M. □

We remark that Theorem 15.3 for $n = 0$ recovers Theorem 12.2 (under the assumption that M is almost geodesically complete).

Appendix A: Alternative formulations of CC^{n-1}

Notation is as in §10.2.

PROPOSITION A1. *The following are equivalent:*

(i) *for any* $m \in \mathbb{Z}$ *and* $-1 \leq p \leq n-1$, *there exists an integer* $\lambda(m) \geq 0$ *such that* $m - \lambda(m) \to \infty$ *as* $m \to \infty$ *and every map* $f : S^p \to X_{(\gamma,m)}$ *extends to a map* $\tilde{f} : B^{p+1} \to X_{(\gamma,m-\lambda(m))}$.

(ii) *for any* $m \in \mathbb{Z}$ *and* $-1 \leq p \leq n-1$, *there exists an integer* $\mu(m) \geq 0$ *such that* $m + \mu(m) \to -\infty$ *as* $m \to -\infty$ *and every map* $f : S^p \to X_{(\gamma,m+\mu(m))}$ *extends to a map* $\tilde{f} : B^{p+1} \to X_{(\gamma,m)}$.

(iii) *for any* $m \in \mathbb{Z}$ *and* $-1 \leq p \leq n-1$ *there exist integers* $\lambda(m), \mu(m) \geq 0$ *such that every map* $f_1 : S^p \to X_{(\gamma,m)}$ *extends to a map* $\tilde{f}_1 : B^{p+1} \to X_{(\gamma,m-\lambda(m))}$, *and every map* $f_2 : S^p \to X_{(\gamma,m+\mu(m))}$ *extends to a map* $\tilde{f}_2 : B^{p+1} \to X_{(\gamma,m)}$.

REMARK. Condition (i) is clearly equivalent to saying that ρ is CC^{n-1} over $e := \gamma(\infty)$. Hence (ii) and (iii) are equivalent formulations of that definition.

REMARK. If $[S^p, Z]$ denotes the set of homotopy classes of maps from S^p to the space Z, then (iii) can be rephrased as saying that the sequence

$$\cdots \leftarrow [S^p, X_{(\gamma,m)}] \leftarrow [S^p, X_{(\gamma,m+1)}] \leftarrow \cdots$$

of sets and functions (induced by inclusions) is "essentially trivial" both as an inverse sequence and as a direct sequence. The terms "pro-trivial" in the inverse case and "ind-trivial" in the direct case are also used.

PROOF OF PROPOSITION A1. (i) $\Rightarrow$ (ii): Given m, p as in (ii) and $\lambda(m)$ as in (i) define $\nu(m) := \min\{r \in \mathbb{Z} \mid r - \lambda(r) \geq m\}$. This is well-defined because $r - \lambda(r) \to \infty$ as $r \to \infty$. Define $\mu(m) := \nu(m) - m$. Then the connectivity condition in (ii) holds. Since $r - \lambda(r) \leq r$, $r - \lambda(r)$ can take on arbitrarily large negative values. Thus $m + \mu(m) = \nu(m) \to -\infty$ as $m \to -\infty$.

(ii) $\Rightarrow$ (i): Given m, p as in (i) and $\mu(m)$ as in (ii) define $\nu(m) := \max\{r \in \mathbb{Z} \mid r + \mu(r) \leq m\}$. This is well-defined because $r + \mu(r) \to \infty$ as $r \to \infty$. Define $\lambda(m) := m - \nu(m)$. Then the connectivity condition in (i) holds. Since $r + \mu(r) \geq r$, $r + \mu(r)$ can take on arbitrarily large positive values. Thus $m - \lambda(m) = \nu(m) \to \infty$ as $m \to \infty$.

(i) $\Rightarrow$ (iii) is clear because (ii) is equivalent to (i) and together (i) and (ii) imply (iii).

(iii) $\Rightarrow$ (i): Let m, p be as in (i) and let $\lambda(m)$ satisfy the first connectivity condition in (iii). Furthermore choose this $\lambda(m)$ so that $m - \lambda(m)$ is as large as possible. Then the sequence $(m - \lambda(m))$ is monotonically non-decreasing. Suppose this sequence were bounded above by q. Let $\mu(m)$ satisfy the second connectivity condition in (iii). If $r > q+1+\mu(q+1)$ then spheres in $X_{(\gamma,r)}$ bound balls in $X_{(\gamma,q+1)}$ so $r - \lambda(r) \geq q+1$ by maximality. This is a contradiction. So $m - \lambda(m) \to \infty$ as $m \to \infty$. □

Note that in the paper it often turns out that Condition (i) holds with constant $\lambda(m) =: \lambda \geq 0$; for example, see the Addendum to Theorem 14.4. The proof of Proposition A1 then shows that $\mu(m)$ in Condition (ii) can also be taken to be constant; in fact $\mu(m) = \lambda$.

Appendix B: Further formulations of CC^{n-1}

Throughout this memoir, including Appendix A, X^n is a cocompact $(n-1)$-connected rigid G-complex with finite (usually trivial) cell stabilizers. There are situations, such as [**BG**], where one wants to replace X^n by a G-space X which is neither cocompact nor a G-complex – perhaps (as in [**BG**]) a non-cocompact G-ANR. It is possible to recognize that a given G-action on the proper CAT(0) space M is CC^{n-1} over M or CC^{n-1} over $e \in \partial M$ by using such a space X, as we now explain.

B.1. Controlled connectivity over points

Let G be a group and let (M, d) be a contractible G-metric space (i.e. G acts on M by isometries). Consider a non-empty G-space X (i.e. G acts on X by homeomorphisms) for which there exists a (continuous) G-map $h : X \to M$. We call such a map a *control function.*

We say that X is *controlled k-connected* (abbrev. CC^k) *over the point $a \in M$* if each cocompact G-subspace K of X lies in a cocompact G-subspace K' such that for each $r \geq 0$ and each $-1 \leq p \leq k$ there exists $\lambda \geq 0$ satisfying:

($*_a$) every singular p-sphere in K over $B_r(a)$ bounds a singular $(p+1)$-ball in K' over[1] $B_{r+\lambda}(a)$.

This property is independent of the choice of control function h, because if h' is another then for each cocompact set $K \subset X$ there is a number δ_K such that $d(h(x), h'(x)) \leq \delta_K$ for all $x \in K$ (compare Proposition 3.1). So only the existence of a control function $X \to M$ is needed for the definition to make sense. We call λ in ($*_a$) a *lag*. In general K' may depend not only on K but also on a, and the lag may depend on K, a, and r. We say that X is CC^k *over M* if both K' and λ are independent of a. Again, this is independent of the control function. By a proof similar to that of Proposition 3.2 we have:

PROPOSITION B1. (i) *X is CC^{-1} over M if and only if the G-metric space M is cocompact.*

(ii) *X is CC^k over M if and only if M is cocompact and X is CC^k over some $a \in M$.* □

These CC-properties of X over M are homotopy invariant in a strong sense:

PROPOSITION B2. **(Invariance Theorem)** *Let $h : X \to M$ and $h' : X' \to M$ be G-maps and let X' be G-homotopically dominated[2] by X. If X is CC^k over a [resp. CC^k over M] then X' is CC^k over a [resp. CC^k over M].*

---

[1]This language is convenient. We will explain it in detail only in this first instance. The condition ($*_a$) says that for every map $f : S^p \to h^{-1}(B_r(a)) \cap K$ there is a map $\bar{f} : B^{p+1} \to h^{-1}(B_{r+\lambda}(a)) \cap K'$ such that $\bar{f} \mid S^p =$ inclusion $\circ f$.

[2]I.e. there are G-maps $\phi : X \to X'$ and $\psi : X' \to X$ such that $\phi \circ \psi$ is G-homotopic to $\mathrm{id}_{X'}$.

The proof is routine and is omitted. □

There is a "best" X to choose over M, namely the universal cover of a $K(G,1)$-complex, by definition a contractible free G-CW complex. Since the action on X is free and M is contractible, control functions $h : X \to M$ exist. This brings us to our main definition: we say that the G-action on M *is* CC^k *over* $a \in M$ [resp. *is* CC^k *over* M] if this X is CC^k over a [resp. is CC^k over M]. This definition is independent of X and h by Proposition B2. More precisely, we can check whether the G-action is CC^k over a, or CC^k over M, by using any contractible space X on which G acts freely and properly discontinuously, and any control map, provided X has the G-homotopy type of a G-CW complex. In particular, X may be the universal cover of a $K(G,1)$-space which is an ANR rather than a CW complex.

PROPOSITION B3. *Let X be a contractible G-ANR, let the action of G on X be properly discontinuous, let $h : X \to M$ be a control function, and let $a \in M$. If G has a torsion-free subgroup of finite index, then X is CC^k over $a \in M$ [resp. is CC^k over M] if and only if the G-action on M is CC^k over a [resp. is CC^k].*

In other words, in this case one can read off the CC-properties of the action on M using X and h.

PROOF. Let X' be the universal cover of a $K(G,1)$-complex, and let G_0 be a torsion free subgroup of finite index in G. The G-space X is CC^k over a if and only if the G_0-space X is CC^k over a, if and only if the G_0-space X' is CC^k over a (by Proposition B2) if and only if the G-space X' is CC^k over a if and only if the G-action on M is CC^k over a. The argument for "CC^k over M" is similar. □

REMARK. In Chapter 3 we defined "the G-action on M is CC^{n-1} over a" or "is CC^{n-1}" only when G is of type F_n. More precisely, we required that the rigid G-complex X on which we measure the CC-property of the action have cocompact n-skeleton. For such a complex X the present definition agrees with the one given there since we may take $K \cap X^n = K' \cap X^n = X^n$ when X^n is cocompact. The new definition is preferable for two reasons: (i) it makes sense even when G is not finitely generated; and (ii) even when G has type F_n it enlarges in a useful way the class of spaces on which the CC-properties of an action can be checked.

B.2. Controlled connectivity over end points

For the rest of this appendix (M,d) is a proper CAT(0) G-metric space, ∂M is its compactifying boundary, and points $e \in \partial M$ are called *end points*. We wish to define "the G-action is CC^k over e". The role of a in §B.1 is played by e here, and the ball $B_r(a)$ is replaced by the horoball $HB_t(\gamma)$ as in Chapter 11.

Let $e \in \partial M$ be represented by the geodesic ray γ. A G-space X with control function $h : X \to M$ is CC^k *over* e if each cocompact G-subspace K of X lies in a cocompact subspace K' such that for each $t \in \mathbb{R}$ and each $-1 \le p \le k$ there exists $\lambda \ge 0$ satisfying

($*_e$) every singular p-sphere in K over[3] $HB_t(\gamma)$ bounds a singular $(p+1)$-ball in K' over $HB_{t-\lambda}(\gamma)$.

and

($**_e$) every singular p-sphere in K over $HB_{t+\lambda}(\gamma)$ bounds a singular $(p+1)$-ball in K' over $HB_t(\gamma)$.

[3]Note that $HB_{t+\lambda}(\gamma) \subseteq HB_t(\gamma) \subseteq HB_{t-\lambda}(\gamma)$.

Of course this[4] depends only on e, not on γ. Moreover, independence of h and the obvious analogue of Proposition B2 concerning G-homotopy invariance hold here too. We say that the G-action on M *is* CC^k *over* $e \in \partial M$ if some (equivalently, any) contractible free[5] G-CW complex X is CC^k over e. If G has a torsion-free subgroup of finite index the obvious analogue of Proposition B3 holds and "the G-action on M being CC^k over e" can be measured using a properly discontinuous G-ANR.

[4]The case $p = -1$ says, in $(*_e)$, that X is non-empty, and, in $(**_e)$, that $h^{-1}(HB_t(\gamma)) \cap K$ is non-empty.

[5]CAT(0) spaces are contractible so the necessary control function exists.

Bibliography

[Å 86] H. Åberg, Bieri-Strebel valuations (of finite rank), *Proc. London Math. Soc.* (3) **52** (1986), 269-304.

[BeBr 97] M. Bestvina and N. Brady, Morse theory and finiteness properties of groups, *Invent. Math.* **129** (1997), 445-470.

[B 81] R. Bieri, *Homological dimension of discrete groups*, QMC Mathematics Notes, 2nc edition. Queen Mary and Westfield College, London.

[B 98] R. Bieri, Finiteness length and connectivity length for groups, *Geometric group theory down under (Canberra, 1996)*, 9-22, de Gruyter, Berlin, 1999.

[BG] R. Bieri and R. Geoghegan, Topological properties of SL_2 actions on the hyperbolic plane, (preprint).

[BG 98] R. Bieri and R. Geoghegan, Kernels of actions on non-positively curved spaces, *Geometry and cohomology in group theory* (P.H. Kropholler, G. Niblo and R. Stöhr, ed.) London Math. Soc. Lecture Notes 252, Cambridge University Press, Cambridge, 1998, 24-38.

[BGr 84] R. Bieri and J.R.J. Groves, The geometry of the set of characters induced by valuations, *J. reine und angew. Math.* **347** (1984), 168-195.

[BNS 87] R. Bieri, W. Neumann and R. Strebel, A geometric invariant of discrete groups, *Invent. Math.* **90** (1987), 451-477.

[BRe 88] R. Bieri and B. Renz, Valuations on free resolutions and higher geometric invariants of groups, *Comment Math. Helvetici* **63** (1988), 464-497.

[BS 78] R. Bieri and R. Strebel, Almost finitely presented soluble groups, *Comment. Math. Helv.* **53** (1978), 258-278.

[BS 80] R. Bieri and R. Strebel, Valuations and finitely presented metabelian groups, *Proc. London Math. Soc.* (3) **41** (1980), 439-464.

[BS 81] R. Bieri and R. Strebel, A geometric invariant for modules over an Abelian group, *J. Reine und angew. Math.* **322** (1981), 170-189.

[BS] R. Bieri and R. Strebel, *Geometric invariants for discrete groups*, (monograph in preparation).

[BrHa] M. Bridson and A. Haefliger, *Metric spaces of non-positive curvature*, Grundlehren der Mathematischen Wissenschaften 319, Springer-Verlag, Berlin, 1999.

[Br 87_{I}] K.S. Brown, Finiteness properties of groups, *J. Pure and Applied Algebra* **44** (1987), 45-75.

[Br 87_{II}] K.S. Brown, Trees, valuations and the Bieri-Neumann-Strebel invariant, *Invent. Math.* **90** (1987), 479-504.

[Bu] K.-U. Bux, Finiteness properties of certain metabelian arithmetic groups in the function field case, *Proc. London Math. Soc. (3)* **75** (1997), 308-322.

[BuGo 99] K.-U. Bux and C. Gonzalez, The Bestvina-Brady construction revisited: geometric computation of the Σ-invariants for right angled Artin groups, *J. London Math. Soc. (2)* **60** 1999, 793-801.

[Du] J. Dugundji, *Topology*, Allyn and Bacon, Boston, 1966.

[Fa 99] F.T. Farrell, Fibered representations, an open condition, *Topology and its Applications* **96** (1999), 185-190.

[FePe 95] S. Ferry and E.K. Pedersen, Epsilon surgery theory, *Proceedings 1993 Oberwolfach conference on Novikov conjectures, rigidity and index theorems*, vol. 2, (A. Ranicki, ed.) London Math. Soc. Lecture Notes, vol. 227, Cambridge University Press (1995), 167-226.

[FrLe 85] D. Fried and R. Lee, Realizing group automorphisms, *Contemp. Math.* **36** (1985), 427-433.

[Ge] R. Geoghegan, *Topological methods in group theory*, (monograph in preparation).

[Geh] R. Gehrke, *Die höhern geometrischen Invarianten für Gruppen mit Kommutatorrelationen*, Dissertation, Frankfurt, 1992.

[Ho 97] P.K. Hotchkiss, The boundary of a Busemann space, *Proc. Amer. Math. Soc.* 125 (1997) 1903-1912.

[Ko] D. Kochloukova, *The FP_m-conjecture for a class of metabelian groups and related topics*, Dissertation, Cambridge, 1998.

[Ko 96] D. Kochloukova, The FP_m-conjecture for a class of metabelian groups, *J. of Algebra* **184** (1996), 1175-1204.

[MMV 98] J. Meier, H. Meinert, and L. VanWyk, Higher generation subgroup sets and the Σ-invariants of graph groups, *Comment. Math. Helv.* **73** (1998), 22-44.

[Me 94] H. Meinert, The geometric invariants of direct products of virtually free groups, *Comment. Math. Helvetici* **69** (1994), 39-48.

[Me 95] H. Meinert, The Bieri-Neumann-Strebel invariant for graph products of groups, *J. Pure Appl. Algebra* **103** (1995), 205-210.

[Me 96] H. Meinert, The homological invariants of metabelian groups of finite Prüfer rank: a proof of the Σ^m-conjecture, *Proc. London Math. Soc.* (3) **72** (1996), 385-424.

[Me 97] H. Meinert, Actions on 2-complexes and the homotopical invariant Σ^2 of a group, *J. Pure Appl. Algebra*, **119** (1997), 297-317.

[Ne 79] W.D. Neumann, Normal subgroups with infinite cyclic quotient, *Math. Sci.* **4** (1979), 143-148.

[On] P. Ontaneda, Cocompact CAT(0) spaces are almost extendible, (preprint).

[Ra] M.S. Ragunathan, *Discrete subgroups of Lie groups*, Ergebnisse der Math., vol. 68, Spring, Berlin, 1972.

[Re 88] B. Renz, *Geometrische Invarianten und Endlichkeitseigenschaften von Gruppen*, Dissertation, Frankfurt (1988).

[Re 89] B. Renz, Geometric invariants and HNN-extensions, *Group Theory (Singapore 1987)*, de Gruyter Verlag, Berlin 1989, 465-484.

[SW 79] G.P. Scott and C.T.C. Wall, Topological methods in group theory, *Homological group theory* (C.T.C. Wall, ed.) London Math. Soc. Lecture Notes 36, Cambridge University Press, Cambridge, 1979, 137-203.

[Se] J.-P. Serre, *Trees*, Springer, Berlin, 1980.

[Sm 57] S. Smale, A Vietoris mapping theorem for homotopy, *Proc. Amer. Math. Soc.* **8** (1957), 604-610.

[Sta 98] C. Stark, Approximate finiteness properties of infinite groups, *Topology* **37** (1998) 1073-1088.

[St 80] U. Stuhler, Homological properties of certain arithmetic groups in the function field case, *Invent. Math.* **57** (1980), 263-281.

[We 60] A. Weil, On discrete subgroups of Lie groups, *Ann. Math.* **72** (1960), 369-384.

[We 64] A. Weil, Remarks on the cohomology of groups, *Ann. Math.* **80** (1964), 149-157.

Editorial Information

To be published in the *Memoirs*, a paper must be correct, new, nontrivial, and significant. Further, it must be well written and of interest to a substantial number of mathematicians. Piecemeal results, such as an inconclusive step toward an unproved major theorem or a minor variation on a known result, are in general not acceptable for publication. Papers appearing in *Memoirs* are generally longer than those appearing in *Transactions*, which shares the same editorial committee.

As of September 30, 2002, the backlog for this journal was approximately 4 volumes. This estimate is the result of dividing the number of manuscripts for this journal in the Providence office that have not yet gone to the printer on the above date by the average number of monographs per volume over the previous twelve months, reduced by the number of volumes published in four months (the time necessary for preparing a volume for the printer). (There are 6 volumes per year, each containing at least 4 numbers.)

A Consent to Publish and Copyright Agreement is required before a paper will be published in the *Memoirs*. After a paper is accepted for publication, the Providence office will send a Consent to Publish and Copyright Agreement to all authors of the paper. By submitting a paper to the *Memoirs*, authors certify that the results have not been submitted to nor are they under consideration for publication by another journal, conference proceedings, or similar publication.

Information for Authors

Memoirs are printed from camera copy fully prepared by the author. This means that the finished book will look exactly like the copy submitted.

The paper must contain a *descriptive title* and an *abstract* that summarizes the article in language suitable for workers in the general field (algebra, analysis, etc.). The *descriptive title* should be short, but informative; useless or vague phrases such as "some remarks about" or "concerning" should be avoided. The *abstract* should be at least one complete sentence, and at most 300 words. Included with the footnotes to the paper should be the 2000 *Mathematics Subject Classification* representing the primary and secondary subjects of the article. The classifications are accessible from `www.ams.org/msc/`. The list of classifications is also available in print starting with the 1999 annual index of *Mathematical Reviews*. The Mathematics Subject Classification footnote may be followed by a list of *key words and phrases* describing the subject matter of the article and taken from it. Journal abbreviations used in bibliographies are listed in the latest *Mathematical Reviews* annual index. The series abbreviations are also accessible from `www.ams.org/publications/`. To help in preparing and verifying references, the AMS offers MR Lookup, a Reference Tool for Linking, at `www.ams.org/mrlookup/`. When the manuscript is submitted, authors should supply the editor with electronic addresses if available. These will be printed after the postal address at the end of the article.

Electronically prepared manuscripts. The AMS encourages electronically prepared manuscripts, with a strong preference for $\mathcal{A}\mathcal{M}\mathcal{S}$-LaTeX. To this end, the Society has prepared $\mathcal{A}\mathcal{M}\mathcal{S}$-LaTeX author packages for each AMS publication. Author packages include instructions for preparing electronic manuscripts, the *AMS Author Handbook*, samples, and a style file that generates the particular design specifications of that publication series. Though $\mathcal{A}\mathcal{M}\mathcal{S}$-LaTeX is the highly preferred format of TeX, author packages are also available in $\mathcal{A}\mathcal{M}\mathcal{S}$-TeX.

Authors may retrieve an author package from e-MATH starting from `www.ams.org/tex/` or via FTP to `ftp.ams.org` (login as `anonymous`, enter username as password, and type `cd pub/author-info`). The *AMS Author Handbook* and the *Instruction Manual* are available in PDF format following the author packages link from `www.ams.org/tex/`. The author package can be obtained free of charge by sending email to `pub@ams.org` (Internet) or from the Publication Division, American Mathematical Society, P.O. Box 6248, Providence, RI 02940-6248. When requesting an author package, please specify AMS-LaTeX or AMS-TeX, Macintosh or IBM (3.5) format, and the publication in which your paper will appear. Please be sure to include your complete mailing address.

Sending electronic files. After acceptance, the source file(s) should be sent to the Providence office (this includes any TeX source file, any graphics files, and the DVI or PostScript file).

Before sending the source file, be sure you have proofread your paper carefully. The files you send must be the EXACT files used to generate the proof copy that was accepted for publication. For all publications, authors are required to send a printed copy of their paper, which exactly matches the copy approved for publication, along with any graphics that will appear in the paper.

TeX files may be submitted by email, FTP, or on diskette. The DVI file(s) and PostScript files should be submitted only by FTP or on diskette unless they are encoded properly to submit through email. (DVI files are binary and PostScript files tend to be very large.)

Electronically prepared manuscripts can be sent via email to `pub-submit@ams.org` (Internet). The subject line of the message should include the publication code to identify it as a Memoir. TeX source files, DVI files, and PostScript files can be transferred over the Internet by FTP to the Internet node `e-math.ams.org` (130.44.1.100).

Electronic graphics. Comprehensive instructions on preparing graphics are available at `www.ams.org/jourhtml/graphics.html`. A few of the major requirements are given here.

Submit files for graphics as EPS (Encapsulated PostScript) files. This includes graphics originated via a graphics application as well as scanned photographs or other computer-generated images. If this is not possible, TIFF files are acceptable as long as they can be opened in Adobe Photoshop or Illustrator. No matter what method was used to produce the graphic, it is necessary to provide a paper copy to the AMS.

Authors using graphics packages for the creation of electronic art should also avoid the use of any lines thinner than 0.5 points in width. Many graphics packages allow the user to specify a "hairline" for a very thin line. Hairlines often look acceptable when proofed on a typical laser printer. However, when produced on a high-resolution laser imagesetter, hairlines become nearly invisible and will be lost entirely in the final printing process.

Screens should be set to values between 15% and 85%. Screens which fall outside of this range are too light or too dark to print correctly. Variations of screens within a graphic should be no less than 10%.

Inquiries. Any inquiries concerning a paper that has been accepted for publication should be sent directly to the Electronic Prepress Department, American Mathematical Society, P. O. Box 6248, Providence, RI 02940-6248.

Editors

This journal is designed particularly for long research papers, normally at least 80 pages in length, and groups of cognate papers in pure and applied mathematics. Papers intended for publication in the *Memoirs* should be addressed to one of the following editors. In principle the Memoirs welcomes electronic submissions, and some of the editors, those whose names appear below with an asterisk (*), have indicated that they prefer them. However, editors reserve the right to request hard copies after papers have been submitted electronically. Authors are advised to make preliminary email inquiries to editors about whether they are likely to be able to handle submissions in a particular electronic form.

TITLES IN THIS SERIES

738 **Yasuro Gon,** Generalized Whittaker functions on $SU(2,2)$ with respect to the Siegel parabolic subgroup, 2002

737 **Arjen Doelman, Robert A. Gardner, and Tasso J. Kaper,** A stability index analysis of 1-D patterns of the Gray-Scott model, 2002

736 **Wojciech Chachólski and Jérôme Scherer,** Homotopy theory of diagrams, 2002

735 **Martina Brück, Xi Du, Joonsang Park, and Chuu-Lian Terng,** The submanifold geometries associated to Grassmannian systems, 2002

734 **Michel Van den Bergh,** Blowing up of non-commutative smooth surfaces, 2001

733 **Milé Krajčevski,** Tilings of the plane, hyperbolic groups and small cancellation conditions, 2001

732 **Jan O. Kleppe, Juan C. Migliore, Rosa Miró-Roig, Uwe Nagel, and Chris Peterson,** Gorenstein liaison, complete intersection liaison invariants and unobstructedness, 2001

731 **Jesús Bastero, Mario Milman, and Francisco J. Ruiz,** On the connection between weighted norm inequalities, commutators and real interpolation, 2001

730 **Suhyoung Choi,** The decomposition and classification of radiant affine 3-manifolds, 2001

729 **Michael Grosser, Eva Farkas, Michael Kunzinger, and Roland Steinbauer,** On the foundations of nonlinear generalized functions I and II, 2001

728 **Laura Smithies,** Equivariant analytic localization of group representations, 2001

727 **Anthony D. Blaom,** A geometric setting for Hamiltonian perturbation theory, 2001

726 **Victor L. Shapiro,** Singular quasilinearity and higher eigenvalues, 2001

725 **Jean-Pierre Rosay and Edgar Lee Stout,** Strong boundary values, analytic functionals, and nonlinear Paley-Wiener theory, 2001

724 **Lisa Carbone,** Non-uniform lattices on uniform trees, 2001

723 **Deborah M. King and John B. Strantzen,** Maximum entropy of cycles of even period, 2001

722 **Hernán Cendra, Jerrold E. Marsden, and Tudor S. Ratiu,** Lagrangian reduction by stages, 2001

721 **Ingrid C. Bauer,** Surfaces with $K^2 = 7$ and $p_g = 4$, 2001

720 **Palle E. T. Jorgensen,** Ruelle operators: Functions which are harmonic with respect to a transfer operator, 2001

719 **Steve Hofmann and John L. Lewis,** The Dirichlet problem for parabolic operators with singular drift terms, 2001

718 **Bernhard Lani-Wayda,** Wandering solutions of delay equations with sine-like feedback, 2001

717 **Ron Brown,** Frobenius groups and classical maximal orders, 2001

716 **John H. Palmieri,** Stable homotopy over the Steenrod algebra, 2001

715 **W. N. Everitt and L. Markus,** Multi-interval linear ordinary boundary value problems and complex symplectic algebra, 2001

714 **Earl Berkson, Jean Bourgain, and Aleksander Pełczynski,** Canonical Sobolev projections of weak type $(1,1)$, 2001

713 **Dorina Mitrea, Marius Mitrea, and Michael Taylor,** Layer potentials, the Hodge Laplacian, and global boundary problems in nonsmooth Riemannian manifolds, 2001

712 **Raúl E. Curto and Woo Young Lee,** Joint hyponormality of Toeplitz pairs, 2001

711 **V. G. Kac, C. Martinez, and E. Zelmanov,** Graded simple Jordan superalgebras of growth one, 2001

For a complete list of titles in this series, visit the AMS Bookstore at **www.ams.org/bookstore/**.